To Connect is to Understand Mathematics 3

Selected works 1999-2007

ALFINIO FLORES

ISBN: 1496150333
ISBN-13: 978-1496150332

DEDICATION

To the memory of Donovan Johnson, who showed me the way to
mathematics education and changed my life.

CONTENTS

ACKNOWLEDGMENTS

The articles were published before in the journals and book chapters indicated. The complete reference to the published version is given in footnotes in the corresponding pages. I appreciate the prompt and free permissions to include these articles in this collection received from the following publishers:

PRIMUS (Taylor and Francis)

Ohio Journal of School Mathematics (Ohio Council of School Mathematics)

*New England Mathematics Journal (*Association of Teachers of Mathematics in New England)

School Science and Mathematics (School Science and Mathematics Association, Wiley)

Convergence (Mathematical Association of America)

International Journal of Mathematical Education in Science and Technology (Taylor and Francis)

Representations and Mathematics Visualization (CINVESTAV)

Contemporary Issues in Technology and Teacher Education

Thanks to the direct intervention of NCTM's President Matt Larson, NCTM gave me permission to use in this collection my articles from the journals *Teaching Children Mathematics*, *Mathematics Teaching in the Middle School*, and the books *Thinking and reasoning with data and chance*, and *Making sense of fractions, ratios, and proportions*.

I am very grateful to my coauthors. It was a pleasure to work with each of them. The following people contributed to some of the articles in this book as coauthors: Vicente Meavilla, Luis Flores, Jeff Samson, Bahadir Yanik, Isabel Perkins, Erin Turner, Renee Bachman, Erika Klein, Jae Baek, Carmina Brittain, Troy Regis, Jon Knaupp, Jim Middleton and Fred Staley. Their names appear in the reference to the published article.

The version of the articles included in this collection stems from my own files. In some cases they do not include the last editorial improvements from the publishers.

1 HISTORY OF MATHEMATICS AND PROBLEM SOLVING: A TEACHING SUGGESTION[1]

We present a teaching suggestion, using the history of mathematics, to give students from different grade levels (from middle school to the university) the possibility to face problems found in old Spanish mathematics books and compare their solutions with those given in those books using the rule of false position and the rule of two false positions.

Key words: Regula falsi; Rule of false position; Rule of two false positions; Tosca; Aurel; Ventallol
2000 Mathematics Subjects Classifications: 01A40, 01A50, 97-03, 97D40

1. The teaching suggestion

Students can gain a better appreciation of the mathematical tools they use, as they compare today's mathematical means with those used in other ages, and an appreciation of the progress of mathematics over time. We also hope that teachers will be attracted by the history of mathematics, consider it as an inexhaustible source of teaching resources, and introduce it in their teaching. The approach can also give students a glimpse on how languages also change over time.

1.1 *The phases*
The teaching suggestion we make has three phases (see Figure 1).

---

[1] Meavilla, V. and Flores, A. (2007). History of mathematics and problem solving: A teaching suggestion. *International Journal of Mathematical Education in Science and Technology*, *38*(2), 253-259. Reprinted by permission of Taylor & Francis (http://www.tandfonline.com).

Phase 1. Phase 1 is the *Analysis of the statement and solving of the problem*. In this first phase, for Spanish speaking students, we give the student a word problem written in old Castilian (Catalan, Galician, …). After analyzing the text (vocabulary, spelling, punctuation, etc.), the statement is adapted to modern language. For students who are not fluent in Spanish, the statement of the problem can be given using the language of the student. Then the student solves the problem using whatever procedure he or she considers appropriate.

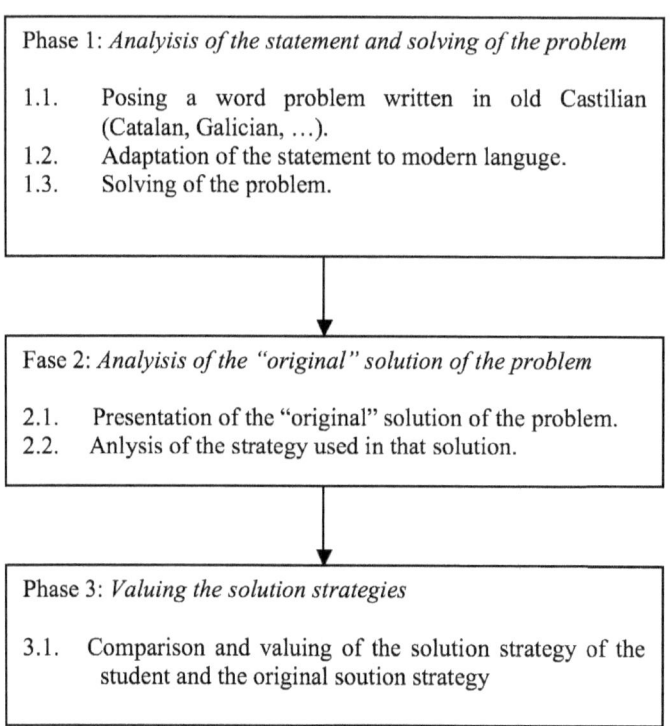

Figure 1. The phases of the teaching suggestion

Phase 2. Phase 2 is the *Analysis of the "original" solution of the problem*, that is, the one given to the statement written in old Castilian (Catalan, Galician, …). The "original" solution is given in modern language for students who do not read the old language, but keeping the old mathematical notation and approach. The student should analyze the strategy used in that solution using appropriate tools.

Phase 3. Phase 3 is *Valuing the strategies of solution*. In this last stage, the students will value his or her solution strategy as compared to the solution procedure analyzed in phase 2.

1.2 *General objectives, learning method, and activities*

Objectives. With the application of the curricular materials in our teaching suggestion we expect students to reach the following objectives
- Recognize and value the usefulness of some old procedures for solving mathematical problems.
- Value the usefulness of the history of mathematics as a source of resources for learning and teaching.
- Accept different approaches to the ones they are used to in the solving of mathematical problems.

Learning method. Work in small cooperative groups (four students per group), and sharing the results of the small groups with the whole group.

Activities for teaching and learning. As an example of the curricular materials that can be used to implement our teaching suggestion, we introduce three activities for teaching and learning geared respectively for students in the middle grades (ages 12 - 16), high school (ages 16-18) and college.

2. Activity 1: The rule of one false position

2.1 *Phase 1: Analysis of the statement and solving of the problem*

Pidese, que el numero 100. se divida en tres partes, que la primera sea dupla de la segunda, y èsta sea tripla de la tercera: que es lo mismo que pedir tres numeros, el primero doblado del segundo, y èste tres doble del tercero, que sumados hagan 100.

It is requested that the number 100 be divided in three parts, that the first be double the second, and this one triple the third: which is that same as asking for three numbers, the first double the second, and this triple the third, that added make 100.

The previous statement is in an 18th century book, *Compendio Mathematico* (1707-1715), written by the Valencian mathematician Thomas Vicente Tosca (1651-1723). Analyze the text by Tosca and adapt it to modern language.

Solve the problem using a procedure you find appropriate.

2.2 *Phase 2: Analysis of the "original" solution of the problem*

Tomo arbitrariamente un numero, y sea 2. èste supongo ser el menor de los tres, que se piden, para mayor facilidad. Triplico el 2. y serà 6. el segundo; duplico el 6. y tengo 12. sumo estos tres numeros 12. 6. 2. y hacen 20. y porque la suma havia de ser 100. busco otro numero por regla de tres, diciendo: Si 20. vienen de 2. de quàntos vendrán 100? Y hallo vienen de 10. Este pues serà el numero menor: luego el segundo es 30. y el mayor es 60. Con esto queda satisfecha la question; porque he dado los tres numeros 60. 30. 10. de los quales 60. es doblado de 30. y èste triplo de 10. y sumados hacen 100.

Thomas Vicente Tosca, *Compendio Mathematico* [1]

I take an arbitrary number and be it 2. This I suppose is the smallest of the three that are asked, for easiness. I triple the 2 and the second will be 6; I double the 6 and I have 12. I add these three numbers and that will make 20, and because the sum had to be 100 I look for another numbers using the rule of three saying: if 20 come from 2, from how much will 100? And I find they come from 10. This will be the smaller number: then the second is 30, and the largest is 60. With this the question is satisfied; because I have given the three numbers 60, 30, 10, from which 60 is double of 30, and this is triple of 10, and added make 100.

The previous text describes the solution of Tosca to the problem posed in phase 1. For the solution, the Valencian author uses a special procedure known as *rule of one false position*, or *simple rule of false position*. Analyze that solution and justify it.

2.3 *Phase 3: Valuing the strategies of solution*

If your strategy and the solution of Tosca are different, compare them and value them. For that you can use the following table 1 (or another similar):

Table 1. Comparison of strategies.

	Your strategy	The strategy of Tosca
Is it general?		
Does it use algebraic symbolism?		
Is it easy to apply?		

3 Suggestions for the instructor

3.1 *Theoretical aspects*

The solution of Tosca is one example of an application of the *rule of one false position*, or *simple rule of false position*, which had been used by the ancient Egyptians, Arabs, and Indians, and was very popular in the mathematical texts of the 15th and 16th centuries. In the words of the Valencian mathematician:

La regla de falsa posición simple se reduce à tres preceptos. 1. Tomese qualquiera numero, que sea apto, para que en èl se puedan exercitar las operaciones que pide la question. 2. Examinese, si es el numero que se pregunta: y si acaso fuere el mismo, quedarà satisfecha la question; pero si no lo fuere, se formarà una regla de tres, que es el tercero precepto, y se hallarà el numero que se busca.

The simple rule of false position is reduced to three precepts. Take any number that is appropriate for applying the operations that are required in the question. 2. Examine if it is the number that is requested: and if it were, the question will be satisfied, but if it were not, the rule of three will be formed, which is the third precept, and the number looked for will be found.

In general the simple rule of false position was used to solve problems of first degree with one unknown, without having to use algebraic symbolism. In fact, the problems solved with the rule of one false position were those whose statements can be translated literally to an equation of the type $a_1x + a_2x + . . . + a_nx = b$ or if you will, $ax = b$.

3.2 *Description of the procedure of Tosca*

The problem in the *Compendio Mathematico*, translated to the language of symbolic algebra becomes the following equation of first degree with one unknown:
$$x + 3x + 6x = 100$$

Let us apply the simple rule of false position to the previous equation. Let us assume that $x = 2$, then:

$$x + 3x + 6x = 2 + 6 + 12 = 20 \neq 100$$

At this point, using the rule of three we can say, "if 20 come from 2, from what will 100 come?" The answer is 10.

Therefore, $x = 10$, $3x = 30$, $6x = 60$ are the requested numbers.

3.3 *Algebraic justification of the rule of one false position*

The equation that solves the problem posed by Tosca is of the type:
$ax = b$ (1)
Let $x = x_1$. Then, $ax_1 = b_1$.
If $b_1 = b$, the solution of (1) is $x = x_1$.
If not, $ax_1 = b_1 \neq b$ (2).
Dividing the equations (1) and (2) term by term we have:

$$\frac{ax}{ax_1} = \frac{b}{b_1} \quad \frac{x}{x_1} = \frac{b}{b_1} \quad x = \frac{bx_1}{b_1}$$

3.4 *Geometric justification of the simple rule of false position*

Solving equation (1) is equivalent to determine the abscissa of point X (see figure 2).

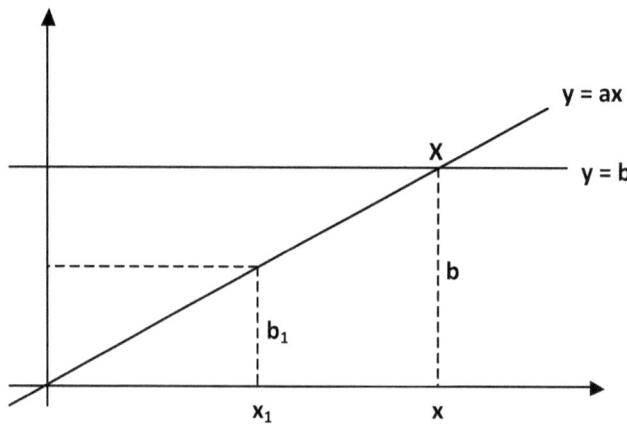

Figure 2. Geometric justification of the rule of false position

By similar triangles we have:
$$\frac{x}{x_1} = \frac{b}{b_1} \quad x = \frac{bx_1}{b_1}$$

3.5 *Mathematical content*

With Activity 1 [*The rule of one false position*] the following contents of conceptual, procedural, or attitudinal nature can be introduced, developed or firmed:

- Algebraic solution of linear problems.
- Manipulation of simple algebraic expressions
- Graphical representation of the functions $y = b$, $y = ax$.
- Graphical interpretation of the equations of first degree with one unknown.
- Graphical interpretation of the solution of an equation of first degree with one unknown.
- Similarity of triangles.
- Search for relations in geometrical figures and configurations.
- Recognition and valuing of the *simple rule of false position* to solve by controlled trial and error some linear problems..
- Flexibility to face problematic situation of algebraic nature with approaches different from the usual.

3.6 *Didactic objective*

Prove that the solution of the equation $ax = b$ is given by the expression $x = \dfrac{bx_1}{b_1}$, where x_1 is an arbitrary value assigned to the unknown and b_1 is the numerical value for the monomial ax for $x = x_1$.

4 Activity 2: The rule of two false positions

4.1 *Phase 1: Analysis of the statement and solving of the problem*

Dos mercaderes lleuan seda, el vno 70 libras: el otro 200 libras: y llegando a vn puerto adonde hauian de pagar cierto derecho: el de las 70 libras, por no tener dineros, da a los del derecho vna libra de seda, y los del derecho le tornan 32 sueldos. Y el de las 200 libras de seda paga de derecho vna libra de seda, y mas 20 sueldos. Demando, a como contaron los del derecho vna libra dela seda, pagando los 2 a vna razon?

Two merchants carry silk, one 70 pounds: the other 200 pounds: and arriving to a port where they had to pay a certain due: the one of the 70 pounds, because he had no money, give to the due one pound of silk, and the ones of the due give him back 32 salaries. And the one with the 200 pounds pays of duty one pound of silk and 20 salaries. I request how did the ones of the duty one pound of the silk, paying the 2 at one rate?

The previous statement is in the book of the 16th century *Libro Primero de*

Arithmetica Algebratica (1552), [First book of algebraic arithmetic] written by the German Marco Aurel, a master living in Valencia. Analyze the text of Aurel and adapt it to the modern language. Solve the problem using the procedure you consider appropriate.

4.2 *Phase 2: Analysis of the "original" solution of the problem*

Pornas por caso y por p.ª falsa posicion que la libra de la seda fue contada en los derechos a 50 sueldos, diras por regla de 3: Si 70 libras de seda pagan de derecho (50 sueldos que es vna libra de seda, menos 32 sueldos que le toman los del derecho, quedan) 18 sueldos, quanto pagaran 200 libras? Hallaras que han de pagar $51\frac{3}{7}$ sueldos, hauian de pagar 70 que es vna libra de seda, y 20 sueldos: vienen de menos $18\frac{4}{7}$ sueldos: así diras (a parte) quando dixe la libra dela seda fue contada a 50 sueldos: hallo $18\frac{4}{7}$ sueldos menos que la verdad. Agora torna a tomar otro qualquiera precio: y sea al presente 80 sueldos la libra. Y diras por la regla de 3. 70 libras de seda pagan 48 sueldos (que es vna libra de seda menos 32 sueldos) 200 libras quanto pagaran? Y hallaras que han de pagar $137\frac{1}{7}$ sueldos hauian de pagar 100 sueldos (que es vna libra de seda, y mas 20 sueldos) paga de mas de lo justo $37\frac{1}{7}$ sueldos. Y tambien diras, quando dixe la libra de la seda fue contada a 80 sueldos: hallo mas de la verdad $37\frac{1}{7}$ sueldos: y pornas los 80 sueldos debaxo delos 50: y los $37\frac{1}{7}$ debaxo delos $18\frac{4}{7}$ y estara en la pratica así.

Posicion p.ª   50   menos $18\frac{4}{7}$ di p.ª

Posicion 2.ª   80   mas $37\frac{1}{7}$ di 2.ª

Las diferencias son $18\frac{4}{7}$, $37\frac{1}{7}$ hazer los has todos de vn genero, y al presente son séptimas, y verna a star como en esta segunda figura veras: no lo digo porque no se pueda hazer con los quebrados, sino por menos fatiga: hazla tu a tu plazer.

D.   50.   menos   130.   A.
B.   80.   mas   260.   C.

Agora summa las dos diferencias: porque son mas y menos, y haran 390. estos seran tu partidor: y multiplica en cruz A con B, y C con D, y summa las 2 multiplicaciones juntas como he dicho: y vernan 23400: los quales partiras por 390: y vernan ala partición 60. A tantos sueldos fue contada la libra dela seda, delos del derecho.

Marco Aurel, *Libro Primero de Arithmetica Algebratica* [2]

Take for case first and by the rule of false position that the pound of silk was counted in the dues as 50 salaries, you will say by the rule of three: If 70 pounds of silk pay of duty (50 salaries that is one pound of silk less 32 salaries, that are received by the duty, remain) 18 salaries, how much did 200 pounds pay? You will find that they have to pay $51\frac{3}{7}$

salaries, they had to pay 70 which is one pound of silk and 20 salaries: they miss $18\frac{4}{7}$ *salaries: so you will say (apart) when I said the pound of silk was counted at* 50 *salaries: I find* $18\frac{4}{7}$ *less than the truth. Now take any other price: and let it be at present 80 salaries the pound. And you say by the rule of 3, 70 pound of silk pay 48 salaries (which is one pound of silk minus 32 salaries) 200 pounds how much will they pay? And you will find that they have to pay* $137\frac{1}{7}$ *salaries, they had to pay 100 salaries (which is a pound of silk plus 20 salaries) pays* $37\frac{1}{7}$ *salaries more than fair. And you will also say when I said the pound of silk was counted at 80 salaries: I find* $37\frac{1}{7}$ *more than the truth: and you will put the 80 below the 50; and the* $37\frac{1}{7}$ *below the* $18\frac{4}{7}$ *and so it will be in practice:*

First position   50   minus $18\frac{4}{7}$ of first

*   Second position   80   plus $37\frac{1}{7}$ of second*

The differences $18\frac{4}{7}$ *and* $37\frac{1}{7}$ *make them all of the same kind, and at present they are sevenths, and they will be as you will see in the second figure: I do not say because it cannot be done with fractions, but for less fatigue: do it as you please.*

D.   50.   *minus*   130.   A.

B.   80.   *plus*   260.   C.

Now add the two differences, and because they are plus and minus, and will make 390, these will be your divisor: and multiply in cross A with B and C with D, and add the two products together as I have said, and it will be 23400: which you will divide by 390 and the division will be 60. So many salaries was one pound of silk counted, by the ones of the duty.

In the previous text we present the solution of Aurel to the problem posed in phase 1. In it the German author uses a special procedure known as the *rule of two false positions* or *rule of double false position*. Analyze this solution and justify it.

4.3 *Phase 3: Valuing the strategies of solution*

If your solution strategy and that of Marco Aurel are different, compare and value them.

5 Suggestions for the teacher

5.1 *Theoretical aspects*

The *rule of two false positions* or *rule of double false position*, apparently of Indian origin, was used preferably to solve equations of first degree with one unknown without using algebraic symbolism. The Arab mathematician al-Amuli [3] (1547-1622) described it in the following terms:

Take for the unknown any number you want: call it first supposition and operate according to the statement. If the equation is verified, that is the unknown. If the balance deviates in more or less, call the difference first deviation. Take another number for the unknown and it will be the second supposition. It will result in general the second supposition. Multiply the first supposition by the second deviation and call the product first result [= R₁]. Then multiply the second supposition by the first deviation and it will give the second result [= R₂]. If both deviation are by excess or both by defect, divide the difference of R₁ and R₂ by the difference of the deviations. In the other case (one by excess and the other by defect), divide the sum of R₁ y R₂ by the sum of the deviations. The quotient will be the number sought.

5.2 *Description of the procedure of Aurel*

Let us follow the procedure of Aurel using the language of symbolic algebra. Let x, expressed in salaries the value of one pound of silk. If 70 pounds of silk pay $x - 32$ salaries, then 200 pounds of silk will pay $\dfrac{200(x-32)}{70} = \dfrac{20(x-32)}{7}$ salaries. On the other hand, as the 200 pounds pay $x + 20$ salaries, we can write the following equation. $\dfrac{200(x-32)}{70} = x + 20$

First supposition

Let us assume that $x_1 = 50$, then: $\dfrac{20(x_1-32)}{7} = \dfrac{360}{7} = 51\dfrac{3}{7} \neq 70 = x_1 + 20$

Let $d_1 = 70 - \dfrac{360}{7} = \dfrac{130}{7} = 18\dfrac{4}{7}$ the first difference (less than).

Second supposition

Suppose that $x_2 = 80$, then:

$\dfrac{20(x_2 - 32)}{7} = \dfrac{960}{7} = 137\dfrac{1}{7} \neq 100 = x_2 + 20$

Let $d_2 = 960/7 - 100 = 260/7 = 37\dfrac{1}{7}$ the second difference (more than).

With this the value of the unknown x is given by the expression:

$$x = \frac{(130/7)\cdot 80 + (260/7)\cdot 50}{(130/7)+(260/7)} = \frac{130\cdot 80 + 260\cdot 50}{130 + 260} = \frac{10400 + 13000}{390} = \frac{23400}{390} = 60$$

5.3 *Algebraic justification of the rule of two false positions*

The equation to solve the problem of Aurel is of the type:

$ax + b = cx + d$ or $(c - a)x + d - b = 0$ (3)

First assumption

Let $x = x_1$

If $ax_1 + b = cx_1 + d$, then $x = x_1$ is the solution of (3)

If $ax_1 + b \neq cx_1 + d$, let $e_1 = cx_1 + d - (ax_1 + b) = (c - a)x_1 + d - b$ (4)

Second assumption

Let $x = x_2$

If $ax_2 + b = cx_2 + d$, then $x = x_2$ is the solution of (3)

If $ax_2 + b \neq cx_2 + d$, let $e_2 = cx_2 + d - (ax_2 + b) = (c - a)x_2 + d - b$ (5)

$(4) - (3) \Rightarrow e_1 = (c - a)x_1 - (c - a)x \Rightarrow e_1 = (c - a)(x_1 - x) \Rightarrow c - a = \dfrac{e_1}{x_1 - x}$ (6)

$(5) - (3) \Rightarrow e_2 = (c - a)x_2 - (c - a)x \Rightarrow e_2 = (c - a)(x_2 - x) \Rightarrow c - a = \dfrac{e_2}{x_2 - x}$ (7)

$(6) = (7) \Rightarrow \dfrac{e_1}{x_1 - x} = \dfrac{e_2}{x_2 - x} \Rightarrow e_1(x_2 - x) = e_2(x_1 - x) \Rightarrow e_1 x_2 - e_1 x = e_2 x_1 - e_2 x$

$\Rightarrow e_1 x_2 - e_2 x_1 = e_1 x - e_2 x \Rightarrow e_1 x_2 - e_2 x_1 = (e_1 - e_2)x \Rightarrow x = \dfrac{e_1 x_2 - e_2 x_1}{e_1 - e_2}$ (8)

Formula (8) deals with all the special cases of the text of al-Amuli and allows us to obtain the value of the unknown x in terms of x_1 (first assumption), x_2 (second assumption), e_1 (first deviation) and e_2 (second deviation).

5.4 *Geometric justification of the rule of double false position*

The solution of the equation $ax + b = cx + d$ (3) is equivalent to the determination of the abscissa of the point of intersection of the graphs of the functions $y = ax + b$ and $y = cx + d$.

Let us assume, without loss of generality, that the graphs of the lines $y = ax + b$ and $y = cx + d$ are like they are described in figure 3

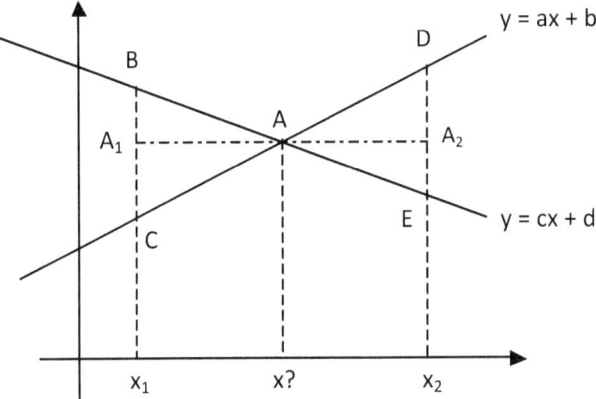

Figure 3. Geometric justification of the rule of two false positions.

Let us assume also that the values x_1 and x_2, assigned to the unknown in the first and second position are such that $0 < x_1 < x < x_2$

- If $x = x_1$ (first supposition), then:

$y_{11} = ax_1 + b$
$y_{12} = cx_1 + d$

    If $y_{11} = y_{12}$, then $x = x_1$ is the solution to the equation (3).
    If $y_{11} \neq y_{12}$, let $BC = y_{12} - y_{11} = e_1$

- If $x = x_2$ (second supposition), then:

$y_{21} = ax_2 + b$
$y_{22} = cx_2 + d$

    If $y_{21} = y_{22}$, then $x = x_2$ is the solution of the equation (3).
    If $y_{21} \neq y_{22}$, let $DE = y_{21} - y_{22} = - e_2$

Because the triangles ABC and ADE are similar we have that:

$$\frac{BC}{DE} = \frac{A_1 A}{AA_2} \Rightarrow \frac{e_1}{-e_2} = \frac{x - x_1}{x_2 - x} \Rightarrow$$

$e_1(x_2 - x) = - e_2(x - x_1) \Rightarrow e_1 x_2 - e_1 x = - e_2 x + e_2 x_1 \Rightarrow e_1 x_2 - e_2 x_1 = e_1 x -$
$e_2 x = (e_1 - e_2)x \Rightarrow x = \frac{e_1 x_2 - e_2 x_1}{e_1 - e_2}$ (8)

5.5 Mathematical contents

With activity 2 [*The rule of two false positions*] one can introduce, develop, or firm the following content of conceptual, procedural, or attitudinal nature:

- Algebraic solution of linear problems.
- Manipulation of simple algebraic expressions
- Graphical representation of functions of the type $y = ax + b$.
- Graphical interpretation of equations of first degree with one unknown.
- Graphical interpretation of the solution of an equation of first degree with one unknown.
- Similarity of triangles.
- Search for relations in geometrical figures and configurations.
- Recognition and valuing of the *rule of double false position* to solve by controlled trial and error some linear problems.
- Flexibility to face problematic situations of algebraic nature with approaches different from the usual.

5.6 Didactical objectives

Show that the solution of any equation of first degree with one unknown can be computed in terms of two arbitrary values assigned to the unknown [x_1 = first assumption and x_2 = second assumption] and of the two corresponding deviations e_1 and e_2.

6 Activity 3: A problem of lendings

6.1 *Phase 1: Analysis of the statement and solving of the problem*

Dos tienen dineros y dice el primero al segundo: si me das 9 de tus ducados yo tendré 3 veces tanto como tú. Responde el segundo y dice: si me das 10 de tus ducados yo tendré cuatro veces tanto como tú. Demando, ¿cuántos ducados tenía cada uno?

Two have money and the first says to the second: if you give me 9 or your ducats I will have three times as much money as you. The second answers: If you give me 10 of your ducats I will have four times as much money as your. I request, how many ducats did each one have?

The previous statement is in the book of the 16th century, *Pratica mercantívol* (1521), written by the Catalonian Joan Ventallol [4]. Analyze the text of Ventallol and adapt it to modern language. Solve the problem using the procedure you consider appropriate.

6.2 *Phase 2: Analysis of the "original" solution of the problem*

De 3 dirás 3/4 y de 4 dirás 4/5, estos quebrados se encuentran en 20. Ahora calcula los 3/4 de veinte y encontrarás que son 15. Después mira cuanto son los 4/5 de 20 y encontrarás que son 16. Suma 15 y 16, hacen 31. Quita 20 de 31, quedan 11 y este es el partidor. Y, dado que el primero dice al segundo que le de 9 ducados de los suyos y el segundo le pide 10, suma 9 y 10, hacen 19. Multiplícalos por los 3/4 de 20, que son 15, hacen 285.

Divídelos por 11 y resultan $25\frac{10}{11}$ de los que debes quitar 9 y quedan $16\frac{10}{11}$. Y tanto tenía el primero. Ahora, para el segundo, toma los 4/5 de 20. que son 16. Multiplícalos por 19 y serán 304, que partirás por 11 y vendrán $27\frac{7}{11}$. De estos quita 10 y quedan $17\frac{7}{11}$. Y tanto tenía el segundo.
Joan Ventallol, *Practica mercantívol* [4]

Of 3 you will say 3/4 and of 4 you will say 4/5, these fractions meet in 20. Now compute the 3/4 of 20 and you will find they are 15. Then look how much are 4/5 of 20 and you will find it is 16. Add 15 and 16, that makes 31. Subtract 20 from 31, there remain 11 and this is the divisor. And given that the first says to the second that he give him 9 ducats of his and the second asks for 10, add 9 and 10, that makes 19. Multiply it by the 3/4 of 20 which are 15, makes 285.

Divide it by 11 and result $25\frac{10}{11}$ *from which you need to subtract 9 and remain* $16\frac{10}{11}$. *And so much had the first one. Now for the second, take the 4/5 of 20 which are 16. Multiply it by 19 and will be 304, which you will divide by 11 and will be* $27\frac{7}{11}$. *From this subtract 10 and* $17\frac{7}{11}$ *remain and that much had the second.*

The previous text describes the solution of Ventallol to the problem posed in phase 1. Analyze this solution and justify it.

6.3 Phase 3: Valuing the strategies of solution

If your solution strategy is different from that of Ventallol, compare and value them.

7. Suggestions for the teacher

The algebraic reconstruction of the procedure of Ventallol can be formulated in the following terms:

$$x + \quad 9 = 3(y - 9) \qquad (9)$$
$$y + 10 = 4(x - 10) \qquad (10)$$

Solving for $y - 9$ from equation (9) we have that:

$$y - 9 = \frac{x+9}{3} \Rightarrow (x + 9) + (y - 9) = (x + 9) + \frac{x+9}{3} = \frac{4}{3}(x + 9) \Rightarrow$$

$$x + y = \frac{4}{3}(x + 9) \Rightarrow x + 9 = \frac{3}{4}(x + y) \quad (11)$$

Solving for $x - 10$ from equation (10) we have:

$$x - 10 = \frac{y+10}{4} \Rightarrow (x - 10) + (y + 10) = \frac{y+10}{4} + (y + 10) = \frac{5}{4}(y + 10) \Rightarrow$$

$$x + y = \frac{5}{4}(y + 10) \Rightarrow y + 10 = \frac{4}{5}(x + y) \qquad (12)$$

Adding (11) and (12), term by term we have:

$$(x + 9) + (y + 10) = \frac{3}{4}(x + y) + \frac{4}{5}(x + y) \Rightarrow x + y + 19 = \frac{31}{20}(x + y) \Rightarrow$$

$$\frac{31}{20}(x + y) - (x + y) = 19 \Rightarrow \frac{11}{20}(x + y) = 19 \Rightarrow x + y = \frac{19 \cdot 20}{11} \quad (13)$$

Substituting (13) in (11) and (13) in (12), we have that:

$$x + 9 = \frac{3}{4} \cdot \frac{19 \cdot 20}{11} \Rightarrow x = \frac{3}{4} \cdot \frac{19 \cdot 20}{11} - 9$$

$$y + 10 = \frac{4}{5} \cdot \frac{19 \cdot 20}{11} \Rightarrow y = \frac{4}{5} \cdot \frac{19 \cdot 20}{11} - 10$$

8. Conclusion

The teaching suggestions we have presented pretends in first place, point to a course of action when using history of mathematics as a didactical resource.

In second term, with the curricular materials we have presented, we hope that the students of the different grade levels become familiar with classical mathematical texts, that without doubt, will help them improve their perception of mathematics.

References

[1] Tosca, T. V. (1707 – 1715). *Compendio Mathematico* (9 vols.). Valencia, A. Bordázar.

[2] Aurel, M. (1552). *Libro Primero de Arithmetica Algebratica*. Valencia, J. Mey.

[3] Sánchez Pérez, J. A. (1949). *La aritmética en Roma, en India y en Arabia*. Madrid, Consejo Superior de Investigaciones científicas.

[4] Ventallol, J. (1521). *Pratica mercantívol*. Publisher and place not given.

2 CALCULUS, PAPER, SCISSORS ✂²

ABSTRACT: The idea of cutting out a curve while looking at the direction of cut of the scissors is used to explore concepts related to curves, tangents and derivatives, especially tangent line, concavity and points of inflection, curvature, and a curve as envelope of tangent lines.

KEYWORDS: Calculus, scissors, tangent line, curvature, envelope, points of inflexion, kinesthetic approach.

INTRODUCTION

The idea of cutting out a curve with scissors can provide a visual and kinesthetic approach to understanding several concepts related to curves, tangents, and derivatives. We also examine direction of a tangent line, concavity and points of inflexion, curvature, and a curve as an envelope of tangent lines.

TANGENT LINES AND CUTS MADE BY SCISSORS

Direction of cut and tangent to a curve

When we cut paper with a pair of scissors, the blades determine a direction

---

² Flores, L. and Flores, A. (2006) Calculus, paper, scissors. *PRIMUS*, *16*(4), 358-362. Reprinted by permission of Taylor & Francis (http://www.tandfonline.com).

when we press them together (Figure 1). When we cut along a straight line, the direction of the scissors remain constant.

Figure 1. Direction of cut.

When we cut along a curve, the direction of the scissors changes as we move along a curve. When cutting at a given point, we can think of the direction of the scissors as the direction of the tangent line of the curve at the point (Figure 2).

Figure 2. Direction of cut and tangent line to a curve.

Concavity and points of inflexion

When students cut out a curve that has a point of inflexion, they will notice that the point where the direction of rotation of the scissors changes corresponds to a point of the curve where the concavity of the curve changes, that is, a point of inflexion. As the curve bends (Figure 3), the direction of the scissors change, in this case first clockwise, and then counter-clockwise. The point when the scissors need to be rotated the other way is the point of inflexion. Another way of thinking about this is to hold the scissors in the same direction and rotate the paper as needed so that the cuts are along the curve. At the point when you need to change the direction of rotation of the paper, that will be the point of inflexion. The concavity of the curve in a region is given by the direction of rotation of the scissors, for example, when concave up, the scissors will be rotating counter-clockwise.

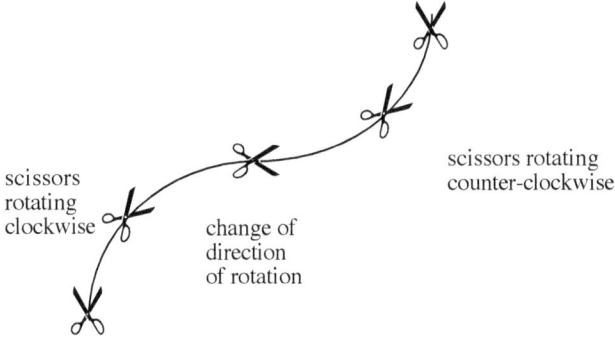

scissors rotating counter-clockwise

scissors rotating clockwise

change of direction of rotation

Figure 3. Point of inflexion.

Curvature

When we cut along a curve, we need to rotate the scissors as we move along the curve. When the curve bends a lot, that is, when the curvature is very big, the scissors need to rotate quite a bit even for small arcs of the curve. For the curve in Figure 4a, the curvature is bigger at *b* than at *a*. The scissors need to rotate more when they are close to *b* than when they are close to *a* for the same amount of arc cut, or equivalently, we need to cut more arc to get the same amount of rotation (Figure 4b).

Figure 4a. Different curvatures.

Figure 4b. Different rates of rotation.

We can relate the change of orientation of the scissors to the corresponding change of direction of the tangent lines at different points of the curve. Let α be the angle of the tangent line with the x axis. We can also compare the change of angle α, $\Delta\alpha$, to the change of arc $\Delta\sigma$ (Figure 5). When the curvature is big, the rate of change $\Delta\alpha/\Delta\sigma$ of the angle to the change in arc length is big. When the curvature is small, that is, when the curve is quite straight, the scissors' rotation is small compared to the length traveled, and thus $\Delta\alpha/\Delta\sigma$ is small. The limit of $\Delta\alpha/\Delta\sigma$ as $\Delta\sigma \to 0$ is $\dfrac{d\alpha}{d\sigma}$. This is the curvature of the curve at the given point. For a more detailed discussion of curvature see [2].

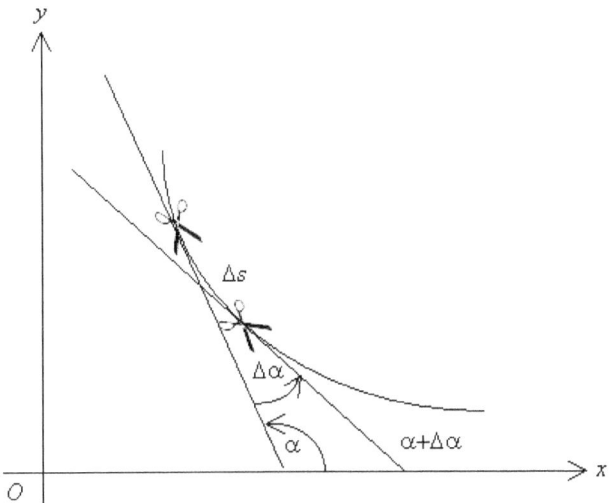

Figure 5. Change of angle and change of arc.

A curve as envelope

When we cut a curve, say an ellipse, we can think of the extended straight cut made by the scissors as being a tangent to the curve (Figure 6). If we consider all the tangent lines, the curve will share one point with each line. Figure 6 consists only of straight lines, and the ellipse would be the envelope, that is, the curve that is tangent to all the straight lines. For more on envelopes, see [1].

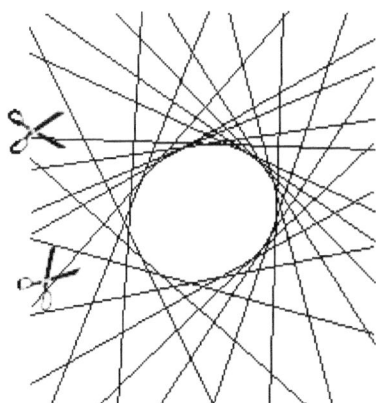

Figure 6. The ellipse being cut out.

REFERENCES

1. Boltianskii, Vladimir. 1964. *Envelopes*. New York, Macmillan.
2. Courant, Richard and Fritz John. 1965. *Introduction to calculus and analysis*.

New York, 1965, pp. 354-360. (Also published by Springer Verlag, 1999).

BIOGRAPHICAL SKETCHES

Luis Flores thought of the idea of using scissors to better understand calculus concepts as a student in high school. In his love for teaching, he likes to develop his own dynamical ways of looking at mathematical concepts. In addition to being an outstanding student, and competing in cross-country, wrestling, and track, he enjoys playing percussion and dancing salsa.

Alfinio Flores teaches mathematics methods courses to present and future teachers. He incorporates many different approaches, such as hands-on materials, visual representations, graphing calculators, computers, and even robots to help students and teachers make sense of mathematics on their own.

3 USING GRAPHING CALCULATORS TO REDRESS BELIEFS IN THE LAW OF SMALL NUMBERS[3]

By the time students are in middle school, they know that with a fair coin the theoretical probability of tossing a head is $1/2$. Students expect that if they repeat the experiment of tossing a coin, the ratio of the number of heads to the total number of tosses will be close to $1/2$. This expectation is justified in the *long* run. The Law of Large Numbers states that in repeated, independent trials with the same probability p of success, the percentage of successes is increasingly likely to be close to the theoretical probability p as the number of trials increases (for a more precise formulation of this law see for example Stark, 2004). However, quite frequently students expect the number of heads and tails to "even out" in the *short* run also. Thus, in a series of coin tosses, many students tend to expect outcomes that alternate frequently between tails and heads. This is a quite common misunderstanding of what it means for the probability of an event to be $1/2$. Some students become uneasy with long runs of heads or tails (even with four or five in a row). This belief that the law of large numbers applies to small numbers as well has been called the belief in the law of small numbers. "Subjects act as if *every* segment of the random sequence must reflect the true proportion" (Tversky and Kahneman 1971, p. 106). It is common for people to believe that a sampling process will be self-correcting. When students are asked to generate a sequence of hypothetical tosses of a fair coin, too often they

---

[3] Flores, A. (2006). Using graphing calculators to redress beliefs in the law of small numbers. In G. Burrill (Ed.), *Thinking and reasoning with data and chance* (pp. 291-304). Reston, VA: National Council of Teachers of Mathematics. Copyright National Council of Teachers of Mathematics. Used by permission.

produce sequences where the proportion of heads in any short segment is much closer to .5 than what would be expected from the laws of chance. Students in the middle grades also know that in the case of a fair die, the probability for each of its faces is 1/6. Here again, a common misconception is that the ratios of the numbers of times faces show up to the total number of trials will "even out" in the short run. All too often students question whether a given die is fair just because in a short run of experiments the same number appears more often than others or because a number fails to appear at all.

Students need to make their conceptions about probability explicit and become fully aware of them. *Principles and Standards* states that "to correct misconceptions, it is useful for students to make predictions and then compare the predictions with actual outcomes" (NCTM, 2000, p. 254). To dispel the belief that coin tosses have the tendency to "even out" in the short run, it is not enough to obtain the overall ratio of heads and tails to the total. It is necessary for students to look at particular sequences of heads and tails that appear and their relative frequencies. Will the sequence HTHT appear more frequently than the sequence HHHH or not? Programmable graphing calculators can be a good tool to simulate random events. With the help of simple programs, students can see in a short amount of time the outcomes of multiple experiments displayed on the screen.

This chapter presents several activities and programs for a graphing calculator that can help students deal with misconceptions about expecting short runs reflect closely the theoretical probability or the long term behavior. First we will present activities that simulate tossing one or two dice, and then activities related to tossing a coin. The programs run on a TI-84 calculator, but other graphing calculators can also be programmed to simulate similar simulations.

Probability programs to simulate tossing dice

The complete set

When students in the middle grades are asked to mentally simulate tossing a die and describe the different outcomes they frequently name all six numbers after only six or seven tosses. While this thinking reflects that each outcome is equally likely, it also reflects that students often expect short runs to reflect closely the theoretical probability. An instructive activity is to have each student actually toss a die and keep a record of how many times it was necessary to toss the die to obtain all six different faces. Students are surprised to find the actual number of tosses needed is quite large. For many students it takes more that 12 tosses, and often one student has to toss the die more than 20 times. After students have done the experiment with real dice they can repeatedly run a program in the graphing calculator and see what happens.

The program ALLFACES simulates tossing a die until all faces have appeared at least once. To encourage student ownership and to get them thinking, have students predict possible outcomes before they toss the die. They could predict how many total tosses will be needed, or how even or uneven the columns will be. How many times will the most frequent number occur before all faces appeared? Let students run the program several times. Figure 1 shows some sample outcomes.

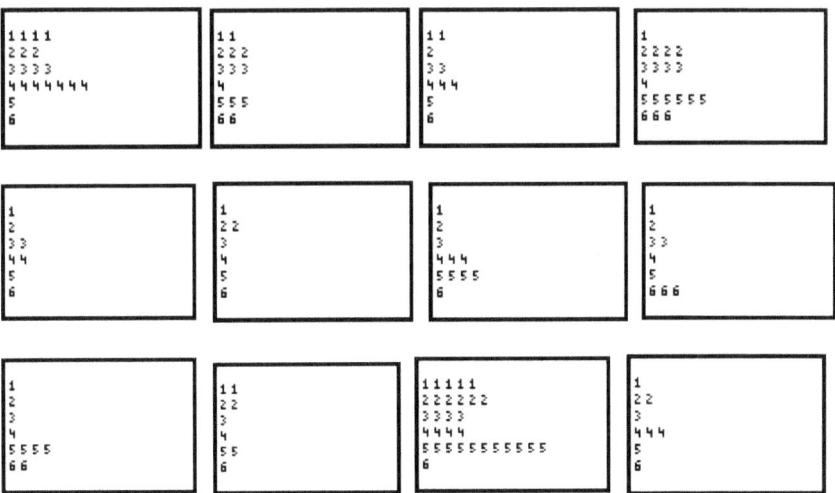

Figure 1. Outcomes for the complete set of faces on a die

Here is the listing of the program. The commands are on the left, and explanations of the purpose of each command are given on the right.
The program can be easily modified to keep track of the total number of times the die was tossed. Students can then graph the total number of tosses over many runs of the program to observe the experimental distribution. They will see that obtaining all six faces after only six tosses is indeed an event that does not happen very frequently. They will also see that in many cases a relatively large number of tosses was needed to obtain the complete set.

PROGRAM:ALLFACES
:6 → dim(L4)   list L4 will have 6 data

```
:Fill (0, L4)              list L4 is filled with zeros
:ClrDraw            clear screen
:Lbl 1                loop starts
:(1 +int (6*rand)) → R   die is tossed
:(1 + L4(R)) → L4(R)        number called advances one
:Text (8R, 6*L4(R)-4, R)  number is displayed on screen at appropriate place
:If min (L4) > 0          checks whether all numbers have appeared once
:Then
:Stop                if they have, program stops
:Else
:Goto 1              if not, goes back to loop
:End                 end of If-Then-Else loop
```

A fair race

Tossing one die can be used to simulate a race. Students can keep track of the cumulative frequencies of each number to see which of the numbers 1 to 6 "wins." After each toss, the number that appears "advances" one place. The first number to appear a given number of times, say six times, will be the winner. For example, if a die is tossed and the first outcomes are 4, 3, 2, 3, then number 3 has "advanced" two places, numbers 4 and 2 one place, and 1, 5, and 6 are still at the starting line. Because all faces have the same probability to show, 1/6, any face has the same probability to win. Will this be a tight race? Students can toss a die first to see what happens in one race. However, to get a better idea of what to expect, many experiments are necessary. Program DIE6 simulates tossing one die until one of the faces shows up six times. Again, before running the program have students predict possible outcomes. Let them run the program several times and then describe in their own words whether they notice anything striking. How does the outcome compare with their predictions? The discussion with students should not focus on predicting the winner but rather on what kind of distribution they might expect to have compared to what the display actually shows, and how the distribution varies with each race. Interesting questions students might consider are how many total tosses are needed before the race is over, how does the number of total tosses vary for each race, how far behind is the last place, or how uneven are the columns. How do the outcomes compare given the fact that each of the numbers 1 to 6 has the same probability of appearing in each toss? Students should notice that although each number has the same chance of advancing, frequently when the winner reaches the crossing line, several numbers are far behind. Have students look at the number with the shortest run, which often will have 0, 1 or 2 tallies. Very seldom will there be a race where the winner is followed closely by two or more numbers. Figure 2 shows some sample outcomes.

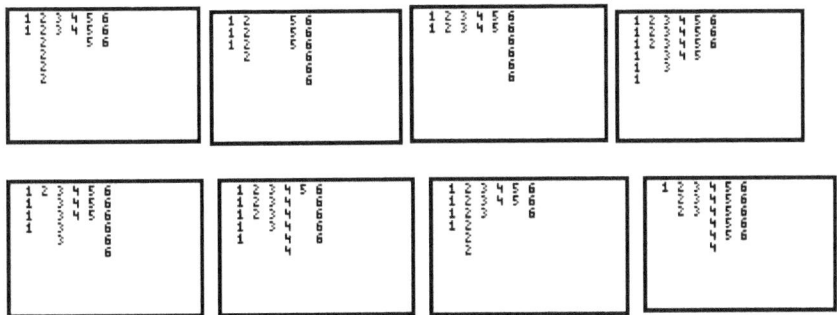

Figure 2. Outcomes of tossing one die

Here is the listing of the program. Commands are on the left, explanations are given to the right.

PROGRAM:DIE6

:6 → dim(L4)	list L4 will have 6 data
:Fill (0, L4)	list L4 is filled with zeros
:ClrDraw	clear screen
:Lbl 1	loop starts
:(1 +int (6*rand)) → R	die is tossed
:(1 + L4(R)) → L4(R)	number called advances one
:Text (6*L4(R)-6, 8R, R)	number is displayed on screen at appropriate place
:If max(L4) = 6	checks whether there is a winner
:Then	
:Stop	if there is, program stops
:Else	
:Goto 1	if not, goes back to loop
:End	end of If-Then-Else loop

They're off!

Students can also gain experience in situations where one outcome has a higher probability than others. A common misconception is that if an outcome has a higher probability of occurring, this will be reflected in the short run also, and the event with the highest probability will occur more often. One activity to address this misconception is to simulate a horse race by tossing two dice and calling the sum (Flores 1990). Again, the frequencies are tallied and every time a sum is called, the corresponding number "advances" one place. After explaining the game, let students pick a favorite number to win and ask them to write down reasons why they picked that number as their favorite. Have students run the race one or two times by

25

actually tossing a pair of dice. Then they can simulate the race on a graphing calculator. The program THEYROFF simulates the tossing of two dice and indicates the sum of the faces. Have students predict a possible outcome, then run the program several times. Have them compare the results on the screen to their predictions and ask whether they would like to choose a different number as the most often sum and why. A sum of 7 has a higher probability of occurring than a sum of 12, because 7 can be obtained from six different outcomes (1+6, 2+5, 3+4, 4+3, 5+2, 6+1), whereas 12 can happen only as 6+6. Does the favorite always win? (See figure 3) Even though a sum of 7 has a higher probability to occur, it is not always the winner.

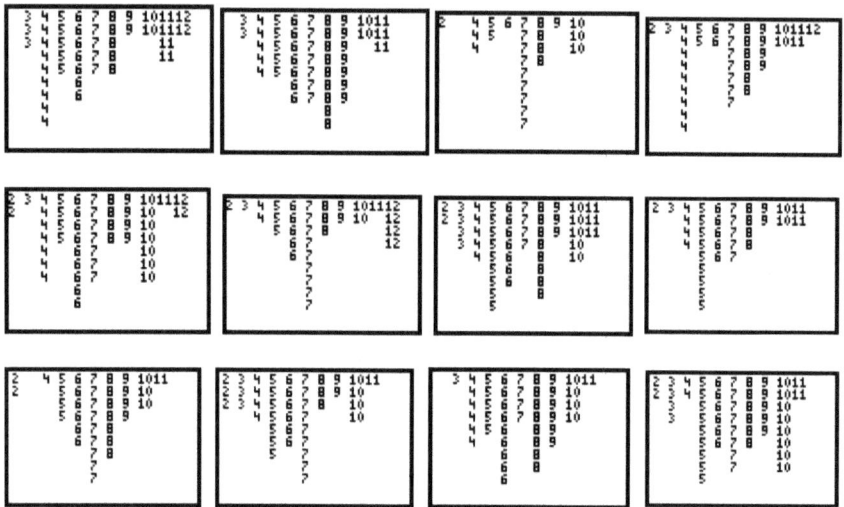

Figure 3. Outcomes of tossing two dice

Here is a listing of the program:

PROGRAM:THEYROFF

:12 → dim(L4)	list L4 will have 12 data
:Fill (0, L4)	list is filled with zeros
:ClrDraw	clear screen
:Lbl 1	loop starts
:(1 +int (6*rand) → R	first die
:(1 + int (6 * rand) → S	second die
:R + S → T	sum is called
:(1 + L4(T)) → L4(T)	number called advances one
:Text (6*L4(T)-6, 8T-14, T)	number is displayed on screen at

26

appropriate place

```
:If max(L4) = 9          checks whether there is a winner
:Then
:Stop                    if there is, program stops
:Else
:Goto 1                  if not, goes back to loop
:End
```

Students can contrast the variability of winners in short races with the outcomes in very long races. In a very long race, say first sum to appear 200 times wins, number 7 will most likely win. The program can be modified to make the race very short (say, by changing the program line **If max(L4) = 9** to **If max(L4) = 3**). Students can see that when the race is shorter it is even harder to predict who will be the winner. Students need to realize that if they toss the dice only a few tens of times, although the numbers in the center tend to advance with more frequency than those on the sides, the distribution will still look rather jaggedly, and not quite like the theoretical distribution (shown in figure 4). Usually a much larger number of tosses is needed to obtain a distribution that closely resembles the theoretical distribution.

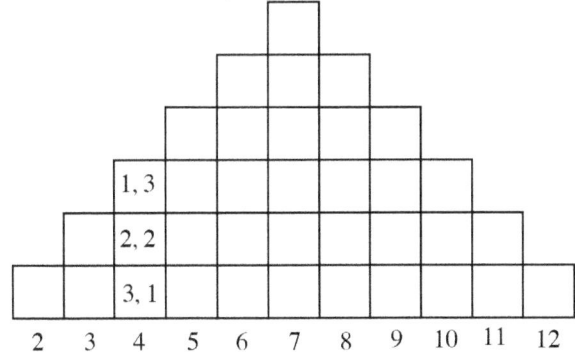

Figure 4. Theoretical distribution of the sum of two dice

[…]

Islands of randomness
The focus of this section is how the tendency to "even out" in the short run is reflected in a hypothetical random table of heads and tails. A random table has a random sequence of H and T in every direction. So, for example, for a ten by ten table, generating one row of ten tosses randomly and then repeating the row ten times would not be a random table, because going down by columns would not generate random sequences of H and T.
In a random table of heads and tails, groups of contiguous heads and groups

of tails appear throughout the table. Long horizontal and vertical runs can frequently occur in a truly random table. A long horizontal run of heads will make it more likely that some of the heads in the row below will be contiguous to some heads above. A larger proportion of longer runs will give rise to bigger contiguous groups. Figure 6 (taken from a random table in Gnanadesikan, Scheaffer, and Swift, 1987) shows a large contiguous group of Hs, some additional Hs and empty spaces corresponding to Ts. Shorter runs of Hs in a row make it less likely that some of the Hs underneath will be part of the same contiguous group. In a table with few long runs most of the contiguous groups of heads or tails are rather small, forming a fragmented pattern.

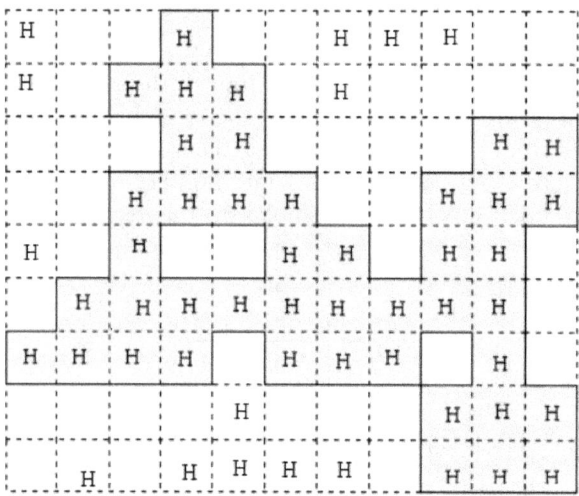

Figure 6. A group of contiguous Hs in a random table

Bringing possible misconceptions to the forefront

Let students generate a 10 by10 random table of heads and tails (a square table with 100 entries in total) by mentally simulating tossing a coin 100 times, and writing the results in a 10 by 10 grid, putting an H in a square if the imaginary toss results in a head, T if it results in a tail. Students could fill the table by rows or by columns or in any other fashion they think will generate a random table. In order to make any preconceptions explicit, it is important that students fill the table of 100 tosses without actually tossing any coins. Students could make their own tables or, as a class, take turns dictating five or six mental outcomes to the teacher who fills in the table on the overhead projector.

H	T	T	H	T	H	T	T	H	H
T	H	H	T	H	T	H	H	T	T

Figure 7. The grid partially filled with hypothetical tosses

Forming islands

In order to make the patterns of Hs and Ts on the table more visual, students can group them into "islands". Two Hs are on the same "island" if their corresponding squares share one side. The common side can be vertical or horizontal. That is, contiguous H squares, either in the same row or the same column, will be in the same island. Two H squares that only share one vertex are not part of the same island unless there is a chain of H squares linking them that share a side pair-wise. T-islands will be formed the same way (see Figure 8). Let students color or highlight all the H-islands.

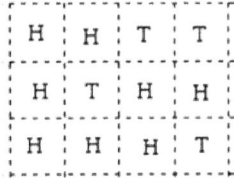

Figure 8. One H island, three T islands

Analyzing the islands

After "tossing" the coins in their heads, students may get something similar to Figure 9. In this example there are 19 H-islands, 23 T-islands, and the biggest island has eight H. There are a total of 52 Hs and 48 Ts (only the H have been labeled, the empty spaces correspond to Ts). Because students have a tendency to "even out" outcomes in the short run, it is very likely their mental simulations will have a fairly large number of separate islands and that the size of the largest island will be relatively small.

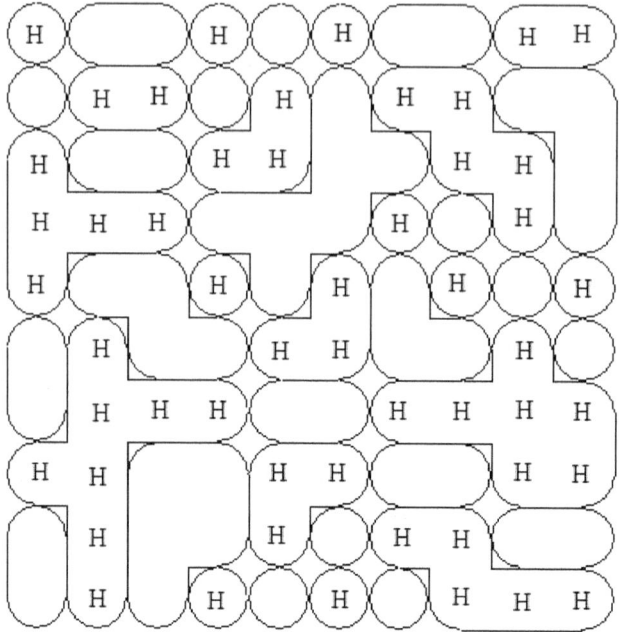

Figure 9. Many small islands

When students believe in the "law of small numbers", it is less likely that a simulation "in their heads" will result in something like Figure 10, where the number of islands is small and the biggest island is quite large (13 H islands, five T islands, maximum size of island is 41; only the Hs have been labeled, empty spots correspond to Ts.

In a truly random experiment, all configurations of heads and tails of the same length have the same probability. A particular configuration that is quite fragmented into many small islands and a particular configuration with large islands have the same probability. Therefore, in a series of random experiments we should also see configurations with big islands, not only very fragmented patterns. However, these big islands are very seldom present in student-generated tables. One reason students do not often generate diagrams with large islands is because they think (consciously or unconsciously) that the number of heads and tails should be about the same, even in the *short* run. So, in their mental simulations they frequently alternate outcomes and often fail to include long runs of the same outcome with the same frequency as they appear in a true random table. The lack of long runs of the same outcome will make it less likely for big islands to form.

Looking at patterns of randomly generated tables and contrasting them with their mentally generated tables can help students confront misconceptions. Graphing calculators can be used to simulate the sequence of one hundred tosses fairly quickly, so that students can look at multiple outcomes and begin to understand the patterns they see or do not see.

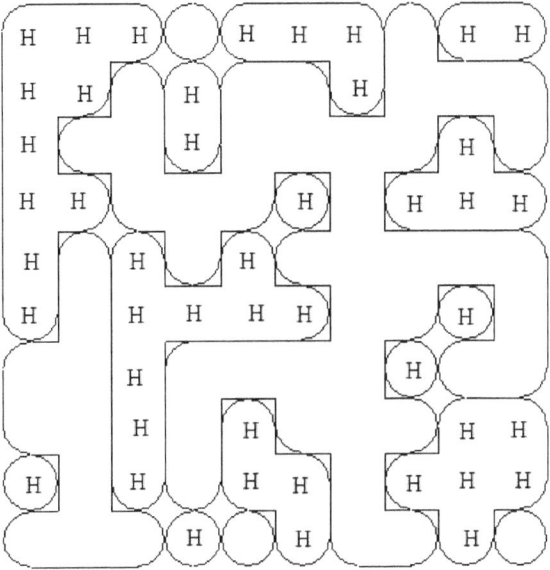

Figure 10. Fewer and bigger islands

Using a graphing calculator

Graphing calculators can be easily programmed to generate random tables. Using a program for the TI-84+ calculator, the tables in Figure 11 were generated. The teacher can generate more tables like these, copy and distribute them in a class or better still, students can use a calculator to generate their own first until they see the diverse outcomes, then be given a copy of a set of tables to color or highlight islands. They will see that very often the size of the largest island is quite big. For example, in Figure 11, the sizes of the biggest islands are as follows: a) 19, b) 36, c) 25, d) 24. The number of islands is accordingly not very large in each case. The total number of islands is a) 28, b) 17, c) 19, and d) 23. Of course, more fragmented patterns will also occur but not as frequently as when students mentally generate the outcomes.

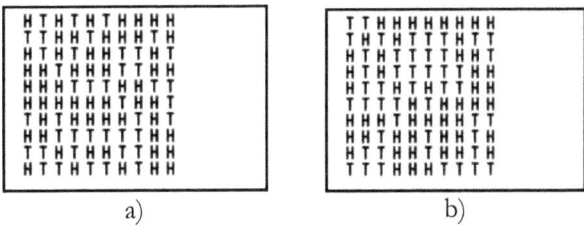

a)   b)

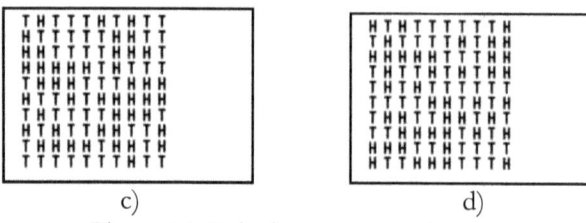

c)   d)

Figure 11. Calculator generated tosses

The program HEADTAIL to generate the tables is given below. It generates ten rows with ten tosses each. A brief explanation of the program is given to the right. By using this program students can generate their own table simulating 100 tosses in a few seconds.

```
PROGRAM:HEADTAIL
:ClrDraw
:For(N,1,10)                    big loop of N rows starts
:For(M,1,10)                    loop for each row starts
:If rand<.5                     start of If... Then... Else loop
:Then
:Text(6*N-6,6*M,"T")           write T if rand < 0.5
:Else
:Text(6*N-6,6*M,"H")           write H if rand ≥ 0.5
:End                            end of If... Then... Else loop
:End                            end of row loop
:End                            end of big loop
```

Of course, the same sequence of 100 outcomes will generate different patterns of islands according to how they are displayed in a table. For example, displaying the 100 tosses in a 20 by 5 table would very likely give a different number for the largest island and for the total number of islands. However, the point is that these displays can be used to help students deal with misconceptions about short term behavior showing that the total number of islands is usually far less than what students expect. This contrast can be done with a 10 by 10 table or with any other array. What is important is the contrast between students' predictions and the actual outcome, not the exact number of islands associated with a particular array.

Extensions
If each student generates a table using a graphing calculator, the class could display in frequency graphs some of the variables associated with the 10 by 10 tables. For example, students could find the size of the largest island in their mental simulation. With the data from everyone in the class, students

can construct a frequency graph of the different sizes of the largest island and see how their mentally generated largest island fits into the random distribution. Students might also consider the total number of islands, locating the total number for their mentally generated table in the frequency distribution of the number of islands generated when using the calculator. Quite often the result of the mental experiment is an outlier. That is, although the result of the mental experiment is on that could happen, it is not of the kind that happens very often. Of course, the islands are only a device to focus the attention on the occurrence of long runs in a true random sequence. That way students will realize that there can be a lot of variability between small samples; and that small samples can give very misleading estimations about the true value of a probability.

Conclusion

Students need to realize that in the short run the ratio of the number of outcomes of a given event to the total number of outcomes does not need to be close to the theoretical probability of the event. In random experiments with probability $1/2$, runs of repeated outcomes occur quite often, so short runs may have many more of one outcome than the other. Outcomes do not always alternate after one or two tosses.

In general, people who believe in the law of small numbers have incorrect intuitions about what to expect from a small sample. They have poor intuitions about confidence intervals, significance, and the power of a statistical test (Tversky and Kahneman 1976). Students need many experiences to develop correct intuitions in probability. Actually tossing coins and dice is important and instructive, but using graphing calculators or computers can provide many more experiences and allow students to conceptualize what they are observing. However, only by making their misconceptions explicit and focusing on what random really means will erroneous conceptions be dismantled. The understanding that it is very likely for the empirical probability to be close to the theoretical probability in the *long* run, but not necessarily after a short number of trials, is a very important step in the development of probabilistic thinking. Students who have a good grasp of this will have a better foundation to understand why the size of the sample is important when trying to make statistical inferences.

This experience of contrasting hypothetical tosses of coins and dice with actual random sequences will also help students understand that it is very difficult to generate random samples mentally. A very common mistake in statistical inference is to assume that a sample is random when in fact it is not. Students will appreciate how tools such as tables of random numbers and calculator or computer generated sequences of random numbers can be useful to obtain a random sample.

References

Flores, Alfinio. They're off!   In *Projects to Enrich School mathematics Level 1*, edited by Judith Trowell (pp. 72-78) National Council of Teachers of Mathematics, 1990.

Gnanadesikan, Mrudulla, Richard L. Scheaffer, and Jim Swift. *The Art and Techniques of Simulation*. Palo Alto, Ca.: Dale Seymour, 1987.

National Council of Teachers of Mathematics. *Principles and Standards for School Mathematics*. Reston, Va.: The Council, 2000.

Stark, Philip B. Glossary of statistical terms, 2004. Retrieved December 20, 2004 from
http://stat-www.berkeley.edu/~stark/SticiGui/Text/gloss.htm

Tversky, A. and Kahneman, D. "Belief in the law of small numbers," *Psychological Bulletin*, *76* (1971), pp. 105-110.

4 HOW DO STUDENTS KNOW WHAT THEY LEARN IN MIDDLE SCHOOL MATHEMATICS IS TRUE?[4]

ACKNOWLEDGEMENT

This article is based on interviews conducted by Sean Berrett, Angie Bilbao, Jamie Bolster, Amelia Clarkson, Hyun Jung Kang, Sarah Kinner, Chris Lemke, Hsiu-Mei Lin, Dmitri Logvinenko, John Lowery, Megan Murphy, Rachel Neuharth, Julia Petersen, Koyal Roy, Jeffrey Samson, Jeanne Scown, Ambur Sedlar, Jane Smoudi, and Sarah Winzeler.

Abstract

This article presents ways in which students ascertain that what they have learned in mathematics is true. Students in the middle school (and a few from other grades) were interviewed by prospective and in-service teachers. Students were asked what they had learned recently in mathematics and how they knew it was true. The answers were grouped by the author according to the justification schemes used by the students in their explanations. Students interviewed used three kinds of justification schemes: externally based, empirical, and analytic. For each kind examples are provided of students' justifications. Additional insights are included from the reflections of the interviewers. Some suggestions are offered regarding how teachers can help increase their students' ability to give convincing arguments in mathematics.

---

[4] Flores, A. (2006) How do students know what they learn in middle school mathematics is true? *School Science and Mathematics*, *106*(3), 124-132. Reprinted by permission of Wiley and School Science and Mathematics Association.

Students in middle school learn a lot of mathematics that goes beyond facts that are obvious or easily verified. Some of the concepts and procedures they learn were developed by mathematicians over several centuries and are by no means evident. How do students know that what they learn in mathematics is true? This article describes ways students in the middle grades justify what they learn in mathematics. The results presented here stem from what some would call an exploratory study rather than a systematic one. However, the data are consistent with results obtained over several years in similar studies. The data presented here are, thus, not isolated anecdotes or outliers, but represent fairly typical and common ways of students justifying what they learned. As such, this study is part of the classroom scholarship devoted to understanding how students come to learn mathematics in a meaningful way. Kyle, a 12-year-old boy in the sixth grade said, "I know it works because my teacher wouldn't lie to me." As Kyle's comment reveals, most students in the middle grades trust their teachers and are convinced that what they are learning is true. Students seldom question the mathematics that is taught in school; many do not develop their own ways to determine why something they learned is true. However, not everything that students think is true, is true. Sometimes they do not remember correctly what the teacher said or they misinterpret the information given. Interviewers in this study reported several misconceptions and misunderstandings. A girl, age 13, in the eighth grade said, "I learned how to do the irrational and rational numbers…. A rational number is a number where it's repeating like .4646464646…. An irrational number can't be made into a decimal…. It's true because my teacher said it was." The truth, of course, is that an irrational number "can be made into a decimal," but the decimal expansion will be infinite and nonperiodic.

It is important for students to develop methods to justify on their own the correctness of a procedure or the truth of a fact. Without these methods, students will not be able to know when they are wrong. Also, as the mathematics becomes more abstract and sophisticated, students need to develop in parallel mathematical sophistication to ascertain that a statement or fact in mathematics is true, or whether a given procedure will give a correct answer. The ways students justify are not always the same as the ways accepted by mathematicians. Teachers need to understand the ways students use to justify so that those methods can be built upon to help students develop their ability to provide convincing arguments in mathematics. At the end of the article some suggestions are provided for teachers of what they can do to increase this ability.

Studies on Proofs and Justifications

Several authors have pointed out the different functions proof has in the teaching of mathematics (Bell, 1976; de Villiers 1999; Hanna & Jahnke 1996). Among the mentioned functions are

- Verification (deals with the truth of a statement)
- Explanation (give understanding of why something is true)
- Systematization (organization of several results in a deductive system of axioms, concepts and main theorems)
- Discovery (invention or discovery of new results)
- Construction of an empirical theory
- Exploration of the meaning of a definition or the consequences of an assumption
- Incorporation of a well-known fact into a new framework and, thus, look at it from a new perspective.

This paper is concerned with the first two functions, verification and explanation. The different functions of proof in mathematics can be emphasized more or less according to the needs of the audience. For example, Hersh (1993) stated that mathematical proof can convince and can explain. He pointed out that in mathematical research the main function of proof is to convince and that at the high school and college level its main function is to explain. Hanna (1990) recommended, whenever possible, the use of proofs that explain rather than proofs that only prove. Both are valid proofs, but a proof that explains, in addition, "must provide a rationale based upon the mathematical ideas involved, the mathematical properties that cause the asserted theorem to be true" (p. 9).

Given the importance of justifications and proofs in mathematics and in the learning of mathematics, it is not surprising that there is a considerable body of research about different aspects related to them. Empirical research in mathematics education has focused, in great part, on describing and analyzing the responses of students to questions that require proof. There is abundant evidence that the majority of students have difficulties in following or constructing deductive arguments presented formally and in understanding how they differ from empirical evidence and how to use them to derive new results (Chazan, 1993; Fischbein, 1982; Harel & Sowder, 1998; Porteous, 1994; Schoenfeld, 1989). Martin and Harel (1989) found that many of these difficulties are also common in prospective elementary teachers.

Research has shown that some students think that measuring, like writing a deductive proof, allows one to reach conclusions that are true and that are applicable to sets that have an infinite number of members. On the other hand, some students see proofs in geometry as proofs for only one case, the case that is represented by the given diagram. They do not appreciate the generic aspects of the diagrams in geometric proofs (Chazan 1993). Schoenfeld (1989) found that students, even after stating that geometric construction and proofs are closely related, behave in construction problems as if their knowledge related to proofs did not exist.

According to Fischbein (1982), when considering the possible impact of a body of information or a pattern of procedures on the dynamic of productive thinking of students who study mathematics, it is necessary to take into account the kind and strength of the credibility attached to them by the student. Learning proofs and learning the general notion of proof is not enough. "The feeling of the universal necessity of a certain property is not reducible to a pure conceptual format. It is a feeling of agreement, a basis of belief, and intuition—but which is congruent with the corresponding formal acceptance" (p. 17). It is important therefore, not only that an argument be correct, but that the student believes and understands it.

Healy and Hoyles (2000) studied the characteristics recognized by 14- and 15-year-old high performing students of the arguments recognized as proofs in algebra, as well as the reasons behind their judgments and the way they construct proofs for themselves. Healy and Hoyles found that students hold two conceptions of proof simultaneously. On one hand, the arguments they consider would receive the best grade, and on the other, those arguments they would adopt for themselves. In the first category the algebraic arguments are popular. In the second, students prefer arguments that they can evaluate and that they find convincing and explaining. These preferences excluded algebraic reasonings. The empirical arguments are most frequent in the proofs constructed by the students, although most of them are aware of their limitations. The most successful students presented the proofs using everyday language instead of using algebra.

Simon (1996) suggested that a student's search for understanding in mathematics and for determining mathematical validity leads not only to inductive and deductive thinking, but also to a third kind, transformational thinking. Although inductive reasoning or deductive thinking can lead students to convince themselves of the truth of a statement, frequently what they search is a sense of how does the mathematical system in question function.

Theoretical Framework: Justification Schemes of Students
This section describes the framework developed by Sowder and Harel (1998) for justification schemes of students. They organized proof (or justification) schemes into three categories: externally based proof schemes, empirical proof schemes, and analytic proof schemes, each with some subcategories

(see Figure 1).

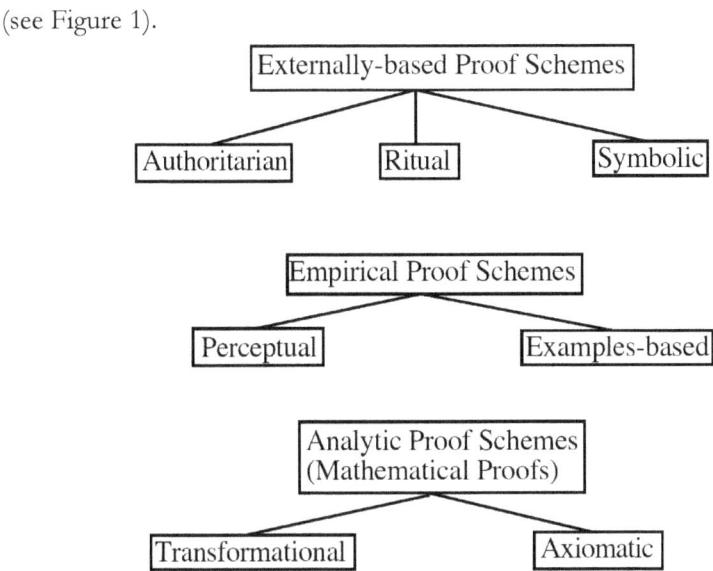

Figure 1. Justification schemes

Externally Based Proof Schemes

What convinces the student and what the student offers to convince others comes from an outside source in the externally based proof schemes. This source can be an authority (the authoritarian proof scheme), the form of the argument (the ritual proof scheme), or meaningless manipulations of symbols (the symbolic proof scheme). Students use the *authoritarian proof scheme* when they rely only on a textbook or a teacher statement to justify a result. When students have these as their only sources for convincing themselves or others, there is cause for concern. A learned helplessness and a rejection of the value of proofs are some the effects of solely relying on authority. A *ritual proof scheme* is shown when a student judges the correctness of an argument solely by the form of the argument rather than by the correctness of the reasoning. Students may doubt that they have given a proof because their argument is not organized in a two-column format or because it does not involve enough mathematical notation, even when the reasoning is sound. The external side of the *symbolic proof scheme* is when symbols are treated as though they have a life independent of any meaning or any relationship to the quantities in the situation where they arise. Examples include writing 6^{12} for $3^3 \times 2^4$, or adding fractions $\frac{1}{2} + \frac{1}{4}$ as $\frac{1+1}{2+4}$. In contrast, a person who is able to unpack correct symbolic reasoning has a transformational proof scheme (discussed later). Thus, when the manipulation of symbols is meaningless, the argument is considered externally based; when meaning underlies the symbolic transformations, the argument is considered analytical.

Empirical Proof Schemes
Justifications based solely on examples exhibit another type of proof scheme. Psychologists have found that most natural human formation of concepts is based on examples and often on just one specific example. Students also appreciate the examples, either as a means of understanding the situation or as a way of checking their understanding. There are two types of empirical proof schemes, the perceptual proof scheme and the examples-based proof scheme. When conclusions are arrived at based on perceptions of a single drawing, or when a drawing is used to convince others, students exhibit the *perceptual proof scheme*. Using one or more examples to convince one-self or others of the truth of a conjecture shows the *examples-based proof scheme*.

Analytic Proof Schemes
Mathematicians consider analytic proof schemes to be the appropriate types of justifications in mathematics. There are two subtypes, transformational analytic proof schemes and axiomatic analytic proof schemes. The general feature of the *transformational proof schemes* is that the justifications of the students deal with the general aspects of a situation and the reasoning is oriented toward proving the conjecture in general.
Mathematicians can organize a developed field of mathematics so that subsequent results follow logically from preceding ones. In the *axiomatic proof schemes*, proofs of new theorems are based solely on axioms or previously proved theorems.

Collection and Analysis of Data
The data were collected through interviews. The interviews were conducted by prospective and in-service teachers as an assignment for a mathematics methods course at Arizona State University. Students in the middle grades were interviewed individually. Some examples from high and upper elementary school students are included, but the topics could correspond to middle school. See Table 1 for a distribution of students by grade level.

Table 1
Distribution of students interviewed by grade level

Grade	5	6	7	8	9	10	11	12
Number of students	2	13	23	20	6	3	2	1

Students were first asked for two facts, procedures, or rules that they had learned recently in mathematics. Then interviewers asked the students how they knew that the fact was true or that the procedure would give a correct answer. Interviewers reported the answers using as much as possible the language used by students. The interviewers also wrote a reflection about the interviews. Some additional insights presented in this article come from these reflections.

The data were analyzed by the author based on the reports of the interviewers. The answers of the students were grouped by the author according to the kind of justification offered by the students. The groupings correspond in a natural way and to a great extent to the schemes of justification identified by Sowder and Harel (1998) for students in high school and college. With some adaptations, the framework was also used before to describe the justification schemes of students in the elementary grades (Flores, 2002).

Schemes of Justification Used by Students in the Middle Grades
Students in the middle school used the three kinds of justification schemes described by Sowder and Harel (1998): (a) externally based proof schemes, (b) empirical proof schemes, and (c) analytical proof schemes. In a few cases, students were not able to provide any justification. For example, Nick, a boy in the sixth grade, stated that any number times 0 is 0. When asked to explain why this was true, Nick replied, "I have no idea."

A student can use different schemes for justification. Many use authority in one case and an empirical or analytical scheme in another case. Some even use different schemes for the same example. For instance, a 13-year-old girl in the eighth grade chose the topic of exponents and was able to explain the meaning of 4^2 and 4^3 in terms of 4×4 and $4 \times 4 \times 4$. When the interviewer asked her about 4^1 she was able to say the answer was 4, saying that you multiply the base only once. However, when asked about the case 4^0, after some initial confusion, the student reverted to the use of authority "anything to the 0 power is always 1." She said that her teacher had told them this a special case.

Externally-based schemes of justification
Students who were interviewed used both authoritarian proof schemes and symbolic proof schemes.

Authoritarian proof schemes. Most students referred to the teacher as the source of authority, and in second place, the textbook used. Although not very frequently, students also used technology as a source of authority. One student justified a procedure because they "did it on the computer." In contrast to younger students, who often cite as sources for mathematical knowledge their parents or older siblings (see Flores, 2002), the mathematical authorities for the middle school students interviewed resided exclusively in school.

In some cases, the students themselves recognized that, even though they trust the authority, they do not have a good understanding of the mathematical situation, as the following quote from Johnny, an 11-year-old sixth grader, illustrates.

> I know this is the right way because my teacher showed us how, and she learned it from her math teacher at the college class, so it came from some really smart people. Other than that I'm not really sure why it makes sense.

A girl, 14 years old, in the eighth grade, explained, "Distance equal speed times the time… The teacher told me the formula and it's written on the book. I never thought about to ask why. I really don't know why it is like that."

In other cases where the justification was based on authority students did not reveal whether they had or did not have an understanding of the situation. A boy, age 12, in Grade 7, said, "When multiplying fractions you do numerator times numerator over denominator times denominator. For example, $\frac{7}{2} \times \frac{14}{3} = \frac{98}{6}$ because $7 \times 14 = 98$ and $2 \times 3 = 6$." When asked how he knew it was true, he said it was how it was taught by the teacher and done in lots of examples. A boy, age 10 in fifth grade described multiplication of fractions. "One half times one third equal one sixth… 1 times 1 is 1, and 2 times 3 is 6. … My teacher taught me so. It's in the book." Hunter, a boy age 14, in Grade 9, explained about solving for the unknown in a proportion. "When a fraction equals another fraction with a variable in it, you cross multiply then solve for the variable." When asked how he knew this is the correct procedure, he said, "Well, it is what my teacher told us to do and I haven't been marked wrong for anything yet."

In several cases, the justification given by students was based on the teachers' authority, but students' answers revealed that they had some conceptual understanding of the situation. Abbi, a girl age 12, in Grade 7, stated, "A rectangle is a parallelogram." Her justification was first based on authority: "My teacher drew us a picture and told us that a rectangle was a special case of a parallelogram." However, when asked why this was true, she was able to explain in conceptual terms. "Well, I guess the sides of a rectangle are parallel just like the parallelogram; they just go straight up and down instead of diagonal."

Symbolic proof schemes. Several students gave justifications that were mainly procedural and symbol driven. This kind of justification was more frequent than in earlier grades where this form of external justification scheme was not used by the children interviewed (Flores, 2002). In many cases, the manipulation of symbols illustrated would correspond to what Skemp (1987) has described as instrumental understanding, or rules without reasons. Students in this section gave sequences of arithmetic procedures as explanations of their thinking, which would correspond to a calculational orientation (Thompson, Philipp, Thompson, & Boyd, 1994). The students focused on computing with the given numbers rather than paying attention to the relations among the quantities in the situation.

Students give sequences of procedures as explanations. Sherman, a 13-year-old boy in the seventh grade, said, "A number times a half is the same as saying that number divided by 2." The interviewer prompted him to explain:

Sherman: I don't know. I just mean … ummmm… like $7 \times \frac{1}{2} = \frac{7}{2}$.

Interviewer: How do you know that this works?

Sherman: Because 7 can be written as $\frac{7}{1}$ and so $7 \times \frac{1}{2}$ is the same as $\frac{7}{1} \times \frac{1}{2}$. Then from there you know you can multiply across the top and across the bottom to get $\frac{7}{1} \times \frac{1}{2} = \frac{7}{2}$

This argument was classified as symbolic because, even though the student clearly knew what he was doing in terms of symbolic manipulation, he did not offer any explanation with respect to the meaning of dividing by 2, or multiplying by 1/2.

Students manipulate symbols with no reference to what the symbols represent. Shelley, a 12-year-old girl in the sixth grade, said, "When you have decimals divided by decimals, you must drag over the decimal point. You cannot divide by anything but a whole number. The number under the dividing sign can be a partial number, but not the divider." When prompted, she explained what dragging meant. "It is when you move the decimal places two places in like this problem, $3.25\overline{)450.07}$. However many points you drag the divider is how many you drag the number under the divider sign." The reason for this dragging she said, was to "increase the value of the number in front." Michelle, a 12-year-old girl in the sixth grade, said she was good at dividing fractions. When given the problem $\frac{3}{4}$ divided by $\frac{1}{4}$, she "flipped" the first number, multiplied the fractions, and got $\frac{1}{3}$. She explained about the flipped part: "That is what you do if you want to get the right answer." Michelle did not show understanding of why a fraction is "flipped." The procedure was merely a symbol manipulation for her. However, when asked what division means, she said it meant, "how many times a number goes into another number," and explained that the problem was asking "how many $\frac{1}{4}$ go into $\frac{3}{4}$." She realized that it would go 3 times and that she had not done the problem right.

Empirical Proof Schemes

The middle school students who used empirical proof schemes went beyond offering simple perceptual arguments. They used examples-based proof scheme. Most of the empirical evidence offered in these interviews falls into two categories, one involving measuring, the other numerical examples.

Examples-based proof scheme. Students justify by measuring or overlapping. Rae, a girl age 16, in the 10th grade, explained about linear pairs of angles.

Rae: Linear pairs add up to 180 degrees
Interviewer: What is a linear pair?
Rae: It's when two angles are right by each other and they form a line. And a line is 180 degrees, so the angles add up to 180 degrees.
Interviewer: Well, how do you know this is true?
Rae: In class we measured a whole bunch of linear pairs to make sure that they added up to 180. And they all did, so I guess they always do.
Rae also had learned about vertical angles: "Vertical angles are congruent.... You can fold the angles and move them around so that they land right on top of each other." Rae drew some vertical angles and folded the paper so that the angles landed on top of each other. "See. They are the same size, so that means that they are congruent."

Students justify generalization by using numerical examples. A girl, age 14, in the 10th grade had learned about the distributive property $a(b + c) = ab + ac$. As justification she said "The distributive property is true because it distributes the *a* into the parentheses. Every time I use different numbers, one side always equals the other side." A girl, age 12, Grade 7, learned a rule to check whether a number is divisible by 3.

Student: To find out if a large number is divisible by 3, you added the numbers, and if that number is divisible by 3, then the original number is divisible by 3.
Interviewer: How do you know that is true for every number?
Student: In class we took a lot of different numbers and added the numbers together to see if it was divisible by 3. To double check, we took the original number and tried to divide by 3. It worked for every number we tried.

Christy, a 13-year-old girl, in the eighth grade, shared that "inverse operations are numbers that undo each other." When prompted to justify, she gave the example, $7 \times 9 = 63$, $63/7 = 9$.

Analytic Proof Schemes
In the analytical proof schemes the arguments are general and the reasoning is mathematical rather than based on examples. In these interviews students used justification schemes that correspond to transformational proof schemes.

Transformational proof schemes. The general characteristic of these schemes is that students are concerned with the general aspects of a situation and involve reasoning oriented toward the general case. Harel and Sowder (1998) pointed out that a particularly important instance of the transformational proof scheme is the transformational use of symbols.

> Symbol manipulations are performed with the intention to derive relevant information that deepens one's understanding… In such activity the individual does not necessarily form specific images for some or all the algebraic expressions and relations… only in critical stages in this process (p. 264).

In several instances students' answers revealed "a rich conception of situations, ideas, and relationships among ideas" which correspond to a conceptual orientation (Thompson et al., 1994, p. 86).

Students use place value and number sense to justify their answers. A 13-year-old girl in the seventh grade first used authority as she quoted the book definition of absolute value: "The absolute value of a number is its distance from zero on a number line." However, the student also showed justification that correspond to transformational schemes. The student used the example of $|-8| = 8$ to explain that both 8 and -8 are the same distance from 0, eight "spaces" to be exact, and since distance is not in terms of negative numbers, then $|-8| = 8$. She also related absolute value to a real world situation. She said that it did not matter if you owe two cookies or if you have two cookies, there are still two cookies involved. Sherman, a 13-year-old boy in the seventh grade explained about addition of decimals. "When you add decimals, you have to line up the decimal points….You write the decimals beneath one another and make sure that the decimal points are right below each other. They have to be in a line." When asked how he knew this is correct he said, Because when you add decimals, you have to make sure that you are adding the right places together. Like… you have to add the ones together, and the tens together and the tenths together, and the hundredths together. So to make sure you do this, you line up the decimals.

Geoff, a 12-year-old boy, had learned to multiply decimals. "It was like when you did a problem, 2.82 × 4, you don't care about the decimal until the end. Out of the top of my head, I think it is 11.22. It is easy for me to do mental math" (the answer was not correct but close). When the interviewer asked why it was not 1.128, Geoff revealed sense of the size of the numbers involved, "What I do is to round 2.83 to 3, 3 × 4 = 12, so the answer is going to be around 12."

Students used operations sense. One example of operation sense was given by a 13-year-old girl, in the seventh grade, as she explained the rule for symbols when multiplying negative numbers. She thought of a negative symbol as another way to say the opposite of something (as in the additive inverse). She explained that, if you had (-6) × 5, you could think of it first as 6 × 5, which is 30, and then take the opposite of 30, which is -30. Then, if both numbers are negative, (-6) × (-5), you would first perform 6 × 5, which is 30, and then take the opposite of 30, which is -30, and then take the opposite of -30, which would then be 30. Abbi, a girl age 12, Grade 7, also revealed operation sense by relating multiplication to repeated addition. The first part of her argument referred to the authority of the teacher, but then she showed operation sense. She stated, "A negative times a positive is always negative.... My teacher said that multiplying is like adding the number to itself. So when you add a negative number to itself it always gets more and more negative." A 12-year-old boy, seventh grade, stated the inverses he had learned so far in school:

Addition is an inverse of subtraction.
Multiplication is an inverse of division.
Subtraction is an inverse of addition.
Division is an inverse of multiplication.

When he thought of inverses in mathematics, he thought about which operations would reverse the process. So if he had to add 9 + 8, in order to get back to 9 he would have to then subtract 8 from the result, that is, 9 + 8 - 8 = 9. The same concepts applied to multiplication and division. Another seventh grade student explained that inverses are mathematical operations that "undo" each other. If you want to go back where you started from, you have to undo the actions you performed.

Students related concepts to previous knowledge. Often students are able to relate concepts to more basic concepts, even when they are not able to justify everything. A 10-year-old boy, in the fifth grade, had learned about area. "The formula for calculating an area of a rectangle is length times width.... I remember the teacher showed us, but I don't remember how the formula came out." However, the same student had a pretty good understanding of the formula of the area of a right triangle.

The area of a triangle is half the base times the height... Let me draw it (he drew a rectangle). Because half of a rectangle is a triangle, so that's half. The base of a triangle is actually the length of a rectangle; the height is the width. Since the area of a rectangle is length times the width, so the area of a triangle is half times the length time width; that's half times the base times the height.

The following justification is from a girl, age 14, in the eighth grade. She used the distributive property and reveals knowledge of the additive inverses.

$a^2 - b^2 = (a + b) (a - b)$. Let me prove it. It's a bit hard. How about backwards? I know how to prove it backwards..... $(a + b) (a - b)$, expand it. It becomes a times a is a^2, then a times $(-b)$ is $-ab$, b times a is ba or ab, b times $-b$ is $-b^2$. So $-ab$ and ab cancel each other, becomes 0.... Just like $-1 + 1 = 0$, so $-ab + ab = 0$. So what is left over is $a^2 - b^2$.

Students used mathematical relations stemming from the situation. Gusto, a boy age 13, in the eighth grade, had a hard time explaining his thoughts. He even mentioned that he always has a hard time when the teacher asked him the same questions during their time together. He explained the process to solve for x in the problem $5 : 6 = x : 36$

First, you have to divide the part of the ratio in the total number, then you times the answer by the other part of the ratio.... I know this is true because the total number of girls is 36, and for every six girls there are five boys. So if there are 36 girls, how many groups is there? First divide six into 36 to find out how many groups of girls there are which is six.

In spite of his difficulties explaining, this student's answer shows that he had a good understanding of the mathematical relations in the situation. His explanation was not merely computational, it revealed the key mathematical issue of the problem, namely proportionality.

Reflections of Interviewers

Several interviewers reported that students have difficulties talking about what they learned in mathematics. One reflected, "I was surprised by the difficulty of students to come up with actual verbal descriptions of what they know in mathematics. The vocabulary and the ability to convey this information were like a foreign language to them." Another interviewer commented that in the interviews "one aspect was common to all: students had difficulty explaining." He also expressed, "It was clear that no one had ever asked these students to explain their understanding of a math fact in words." Some evidence that students are not usually asked to explain was offered by Christy, a 13-year-old girl in the eighth grade. She mentioned that she knew a lot of the facts and rules of mathematics, but she did not know why they were true. She also stated. "This is the first year that anyone has ever asked me why I am doing anything!"

This difficulty in explaining should not be surprising if teachers themselves do not often explain why something is true. Anthony, a boy age 13, in the eighth grade, shared, "When you multiply two negatives together you get a positive." He said that was just the way he was taught. The interviewer asked him how he knew that it worked and he responded that it was never explained to him. He said that they do not give a reason behind why it works, they do not go in-depth, they just say it works. Another student said "The teacher doesn't teach us why it works in school. He just says here is the formula or equation and tells us to use it."

Several students also had difficulty in remembering what they had learned recently. One interviewer remarked that on top of everything else, he was amazed at how little students were able to say even what they learned this year. Some students perceived the object of mathematics class was to compute something or find an answer rather than to learn or remember something. The following dialog is with a 10-year-old boy in the fifth grade. When asked what he had learned in mathematics, after a pause of about 30 seconds, he finally responded,

Student: I really don't know. You didn't give me the problem or the number to solve.

Interviewer: What I mean is a rule or formula, for example, odd number plus even number is odd number.

Student: Well, I get it, but it is still too hard. I don't remember any.

Interviewer: How about the facts you have learned last year?

Student: If I don't remember anything this year, how can I remember last year?

Many students became distrustful about being asked how they knew something was true. One interviewer reflected,

> All of the students were surprised when I asked them how they knew what they had learned was true. They acted suspicious, like I was asking them a trick question. This in and of itself was revealing. Obviously students are used to being handed information and expected to believe it without question.

Another interviewer remarked, "The idea of 'proving' one's answer was foreign to all of the children." An interviewer wrote in his reflection: "The second question (i.e., how do you know it is true) really stops students in their tracks." He described how a girl's behavior changed when he asked the second question: "She was smiling and moving around up to this point and when I asked the question the smile vanished and she stopped moving."

Daniel, a 12-year-old seventh grader, was explaining about coordinate axes. When asked the second question, he exclaimed, "What? You're confusing me. Are you trying to mess with me?"

In their final reflections two interviewers highlighted the value of the interviews for learning about students' thinking and how they could incorporate them to improve their teaching. Describing the variety of ways used by students to justify their knowledge, one prospective teacher realized how important it was for her as a teacher to try to explain concepts in a variety of ways. She also realized that not all students learn by providing mathematical explanations, but that students can learn mathematical concepts by relating real world scenarios to them. Another interviewer, an experienced teacher, stated that she was amazed how much she learned about each of the students during the one-on-one interviews, and what each one needed. She became aware of the importance of interviews as a means of assessment. The interview process also helped her set her own "goals of giving students the opportunity to move from concrete to abstract."

Final Remarks

The examples presented here are consistent with data gathered along several years through interviews conducted by prospective and in-service teachers. Year after year, interviewers find that a fairly large proportion of the students use mainly authority to justify what they have learned. A fairly large number of students use an empirical approach, in many cases using only one example to ascertain truth of a mathematical statement. Few students spontaneously offer analytical explanations. Year after year, interviewers report that students have difficulty explaining what they have learned in mathematics. However, also in a consistent way there are students who are quite capable, after some probing and given enough time, to explain in meaningful ways the mathematics they are learning.

Teachers can help students develop proof schemes by modeling themselves what they should expect from students. A teacher can justify formulas and procedures by explaining the reasons of why these work; explain why statements are true, or why concepts or symbols are defined the way they are. Students will in this way relate notation to the meaning behind it. For example, Shelly, the student who explained how to solve $3.25\overline{)450.07}$ in terms of dragging the decimal point, was very pleased to understand that by multiplying both the divisor and the dividend by 100 the problem would be transformed into $325\overline{)45007}$, and the answer would be the same, an analogous situation to multiplying a fraction by $\frac{100}{100}$.

49

Teachers can also help students develop their ability to talk about mathematical ideas by giving them the opportunity to share and explain their thinking to their classmates and to the teacher. Sometimes this is hard to do, as one of the interviewers remarked: "I found it hard for myself to not to say anything or guide them in giving an explanation." However, after giving students the opportunity to think about the rule, they were "able to explain how they knew a particular rule was true."

Very importantly, teachers can help students develop their proof abilities by asking them to justify their answers, both when they are correct and when not. Students need to get used to questioning and explaining their own answers so that, gradually, they will think about what is presented to them and not just learn the new content with blind faith. The practice of justifying in class will also help students remember what they are supposed to learn by highlighting what are the important ideas in a lesson, the concepts on which the exercises and computations are based, and the connections of these concepts with previously learned concepts.

References

Bell, A. W. (1976). A study of pupils' proof-explanations in mathematical situations. *Educational Studies in Mathematics, 7,* 23-40.

Chazan, D. (1993). High school geometry students' justifications for their views of empirical evidence and mathematical proof. *Educational Studies in Mathematics, 24,* 359-387.

de Villiers, M. D. (1999). *Rethinking proof with the Geometer's Sketchpad.* Emeryville, CA: Key Curriculum Press.

Fischbein, E. (1982). Intuition and proof. *For the Learning of Mathematics, 3*(2), 9-18, 24.

Flores, A. (2002). How do children know what they learn in mathematics is true? *Teaching Children Mathematics, 8,* 269-274.

Hanna, G. (1990). Some pedagogical aspects of proof. *Interchange, 21*(1), 6-13.

Hanna, G., & Jahnke, H. N. (1996). Proof and proofing. In A. J. Bishop, K. Clements, C. Keitel, J. Kilpatrick & C. Laborde (Eds.), *International handbook of mathematics education* (Vol. 2, pp. 877-908). Dordrecht: Kluwer Academic Publishers.

Harel, G., & Sowder, L. (1998). Students' proof schemes: Results from exploratory studies. In A. H. Schoenfeld, J. J. Kaput & E. Dubinsky (Eds.), *Research in collegiate mathematics education. 3* (pp. 234-283). Providence, RI: American Mathematical Society.

Healy, L., & Hoyles, C. (2000). A study of proof conceptions in algebra. *Journal for Research in Mathematics Education, 31*(4), 396-428.

Hersh, R. (1993). Proving is convincing and explaining. *Educational Studies in*

Mathematics, 24(4), 389-399.

Martin, W. G., & Harel, G. (1989). Proof frames of preservice elementary teachers. *Journal for Research in Mathematics Education, 20*(1), 41-51.

Porteous, K. (1994). When truth is seen to be necessary. *Mathematics in School, 23*(5), 2-5.

Schoenfeld, A. H. (1989). Explorations of students' mathematical beliefs and behavior. *Journal for Research in Mathematics Education, 20,* 338-355.

Simon, M. A. (1996). Beyond inductive and deductive reasoning: The search of a sense of knowing. *Educational Studies in Mathematics, 30,* 197-210.

Skemp, R. R. (1987). Relational understanding and instrumental understanding. In *The psychology of learning mathematics* (pp. 152-163). Hillsdale, NJ: Erlbaum.

Sowder, L., & Harel, G. (1998). Types of students' justifications. *Mathematics Teacher, 91,* 670-675.

Thompson, A. G., Philipp, R. A., Thompson, P. W., & Boyd, B. A. (1994). Calculational and conceptual orientations in teaching mathematics. In D. B. Aichele (Ed.), *Professional development for teachers of mathematics* (pp. 79-92). Reston, VA: National Council of Teachers of Mathematics.

5 QUOTIENT AND MEASUREMENT INTERPRETATIONS OF RATIONAL NUMBERS[5]

Teaching rational number concepts is a challenging task for most teachers. National assessments show that many students in the U. S. have a poor understanding of rational numbers. For example, only 50% of 4th graders could answer correctly how many fourths make a whole, only 45% could identify a diagram illustrating the equivalence of two fractions, only 21% could compare fractional representations, and only 10% could show how three figures can be divided (Wearne and Kouba 2000; Arbaugh, Brown, Lynch, and McGraw 2004).

The part-whole relation is the most used model for teaching fractions. In this approach an object or a set is partitioned into equal-sized parts and those parts are compared with the whole unit. Understanding fractional parts as equal-sized and the corresponding notation (numerator, denominator) are some of the important ideas in this model. Often fraction instruction relies solely on the part-whole interpretation of fractions. However, research has shown that the part-whole interpretation of fractions is not sufficient as a foundation for the system of rational numbers (Lamon, 2001). Rational numbers are very complex, and teaching just the part-whole interpretation does not generate a complete understanding.

There are various meanings of rational numbers besides part-whole, these include: rational numbers as quotients, as measures, as ratios, and as

---

[5] Flores, A., Samson, J., Yanik, B. H. (2006). Quotient and measurement interpretations of rational numbers. *Teaching Children Mathematics*, *13*(1), 34-39.

operators. Seven pizzas divided among four people is an example of a quotient context. Seven quarters of flour is an example of measure interpretation. Stating that there are 7 female teachers for every 4 male teachers in a school district is an example of a ratio. Thinking about 7/4 as shrinking by a factor of 4 followed by a stretching by a factor of 7 is an example of an operator. Students who work with these interpretations of rational numbers and have time to explore all of them will develop a more structured and usable knowledge base of the rational numbers than students who spend all their time only with the part-whole model.

In this article we will give some ideas for teaching two of the additional interpretations of fractions, the quotient and measurement models. As with other interpretations of rational numbers, quotient and measurement are best learned by following these principles based on research and draw on experience (Cramer, Behr, Post, and Lesh 1997, p. 2):

(a) Children learn best through active involvement with multiple concrete models,

(b) Physical aids are just one component in the acquisition of concepts: verbal, pictorial, symbolic and real world representations also are important,

(c) Children should have opportunities to talk together and with their teacher about mathematical ideas, and

(d) Curriculum must focus on the development of conceptual knowledge prior to formal work with symbols and algorithms.

Quotient interpretation of rational numbers
Building a foundation for a solid understanding of rational numbers as quotient starts in the early grades (Lamon 1999). The quotient understanding of fractions is based on the idea that a fraction may be viewed as the result of a division. For example, if three apples are shared by six people, each person gets 3/6 of an apple. The same symbols are used in both the context of division and of fractions. Often a slanted bar (/) is used to represent fractions, and on some calculators it is used to label the division key. However, we should not assume that children will immediately recognize that we can move from one context to the other. Some children are surprised to see that the result of a division like $7 \div 4$ is a fraction where exactly the same digits appear as numerator and denominator (Flores and Klein 2005). Some children will object to using fractions when the context is about division (Toluk 2001), especially if they have learned division in contexts where subdividing the unit does not make much sense. For example, based on experience, 3/4 of a marble is not a very practical answer. Behr, Harel, Post, and Lesh (1992) suggest that students need to consider both continuous and discrete objects and different possible ways of forming units for the

53

numerator and denominator to see a complete picture of a rational numbers as a quotient.

Partitioning activities are a good place to start (as they are in the part-whole relations) for understanding rational numbers as a quotient. Moreover, when we draw on student strategies in activities that support fair sharing students develop more flexible thinking (Empson 2002). However, the teacher needs to keep in mind that the interpretation of the concept of partitioning is slightly different than its interpretation in part-whole situations in that the quotient interpretation is closely related to characteristics of division (de Silva, 2001). In the quotient situation, the dividend and the divisor represent typically different things (e.g., 7 pizzas shared by 4 people). Lamon (1999) suggests asking questions to emphasize the quotient perspective, such as, How much is *one share*? Or, How much does *one person receive*? To emphasize the part-whole relations, one may ask, What part of the pizza will each person receive?

Partitioning involves two types of division: partitive (sharing) and measurement (servings) division. The partitive interpretation of division involves fair sharing situations. For example, for the division $7 \div 4$, the dividend 7 represents the size of a quantity (unit), and the divisor 4 represents the number of groups or parts. The fraction 7/4 represents the size of each part in combination. In the measurement interpretation of division, the divisor represents the amount in each group, which is the size of the measuring unit. Measurement division problems are not asked as frequently as the partitive division problems (Lamon 1999). An example of measurement division problem would be: If a batch of cookies require 6 ounces of butter, how many batches can I cook with 3 ounces of butter. One dilemma faced by teachers in the teaching division is that not all children will realize this difference, and, if they do not, teachers need to consider when to draw their attention to it.

According to Behr, Harel, Post, and Lesh (1992), there are two types of rational number that result from a partitive interpretation of division. One type is related to single unit and the other is related to composite units. Consider the example: 3 pizzas shared by 4 people. In the first interpretation, the unit is a pizza – a single unit. In this case, each person gets one fourth of each pizza (total three fourths) (Figure 1). In the second interpretation, three pizzas are considered as a composite unit. In this situation, each person gets $\frac{1}{4}$(3-unit) pizza (Figure 2). There are several ways of representing the equal parts that each person receives, but all persons get the same amount of pizza at the end.

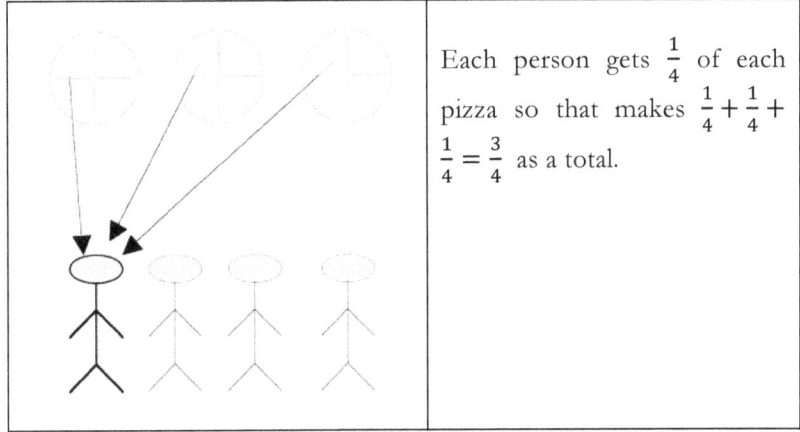

Each person gets $\frac{1}{4}$ of each pizza so that makes $\frac{1}{4} + \frac{1}{4} + \frac{1}{4} = \frac{3}{4}$ as a total.

Figure 1. Three one fourths

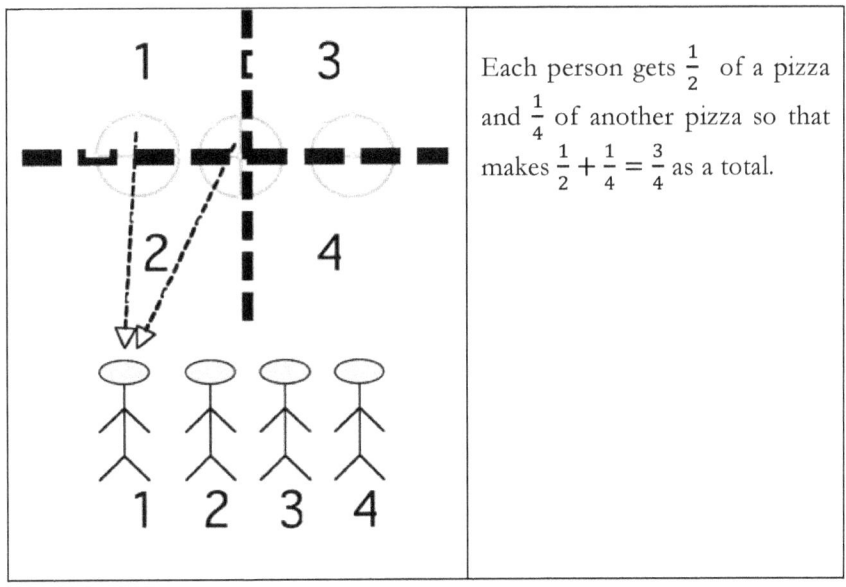

Each person gets $\frac{1}{2}$ of a pizza and $\frac{1}{4}$ of another pizza so that makes $\frac{1}{2} + \frac{1}{4} = \frac{3}{4}$ as a total.

Figure 2. One fourth of 3.

The quotient perspective is also important to make a connection between fractions and decimal representations of rational numbers, either by using the long division algorithm or the division key in a calculator (Figure 3). Work on the quotient perspective helps reinforce the connection via equivalent fractions, such as $3/4 = 75/100$. Dealing with the result of the division will also help students begin to see a fraction as a single number.

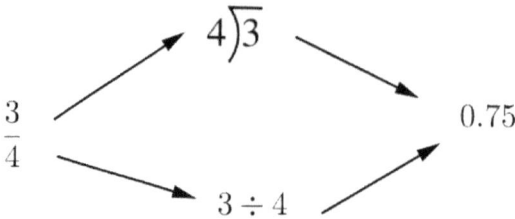

Figure 3. Fractions, division, and decimals.

Thinking about fractions as divisions can also help students develop more flexibility. Heidi was asked to find a fraction between 1/5 and 1/4, She immediately wrote $\frac{1}{4.5}$ (she then wrote the equivalent $\frac{2}{9}$). Heidi used the inverse relation between the size of the divisor and the result of a division (keeping the dividend constant) to find the intermediate fraction. It would be hard to think about $\frac{1}{4.5}$ using part-whole model.

Measure interpretation of rational numbers

Real life experiences where measurements are required are a good starting point for instruction on the measure interpretation of rational numbers. Discussions of these situations lead to the main ideas that go along with this topic: 1) accuracy of measure, 2) density of rational numbers, 3) successive partitions, and 4) fraction equivalence. The measurement interpretation relies on two principles of measurement. One is the inverse relation between the size of the measuring unit and the number of times used in the measure of a given quantity. The other is the possibility of partitioning the unit into smaller and smaller units until one can approximate a given quantity with any desired precision.

Finding equivalent fractions using fraction strips or fraction bars can help establish a bridge between part-whole and measurement on the number line. The use of a number line is a great way to discuss measurement. Students have experience with number lines and with standard and non-standard measurement from a young age. Finding common benchmarks, such as 1/2, 1/4, and 3/4 on a number line that can be obtained by halving is a good way to make the transition into rational number measurement.

We should not, however, assume that students will immediately incorporate all the aspects necessary to accurately represent fractions on the number line.

Figure 4 shows Abigail's effort to represent 1/5 and 1/4 on a number line (Rothery 2005). While her diagram correctly shows that they are bigger than 0 and smaller than 1, and that 1/5 is smaller than 1/4, the distances between the fractions are not accurately represented. The distance between 0 and 1/4 should be one-fourth of the distance from 0 to 1. The distance between 1/5 and 1/4 should be much smaller than that between 0 and 1/5.

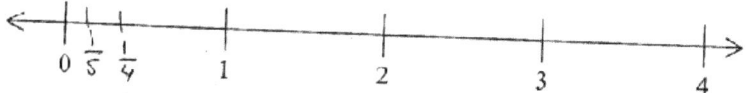

Figure 4. Abigail's representation of 1/5 and 1/4 on the number line.

Students will not automatically pay attention to the same features of numbers on a number line as teachers do. With the help of a computer program, Abigail obtained a graph representing 1/5, 1/4, 1/3 and 1/2 (Figure 5). Abigail was asked by her teacher to reconstruct the graph by hand. Abigail correctly changed the decimals back to common fractions and correctly represented the order among them (Figure 6). However, she did not seem to pay attention to the distances between the points on the computer display. Her graph was not a result of a hasty drawing. In her graph the distances between 1/5 and 1/4, between 1/4 and 1/3, and between 1/3 and 1/2 are almost the same length. Abigail's representation reveals that students do not always internalize properties of numbers, such as distances between them, just by looking at a picture or a representation on computer screen.

Figure 5. Computer display of 1/5, 1/4, 1/3 and 1/2

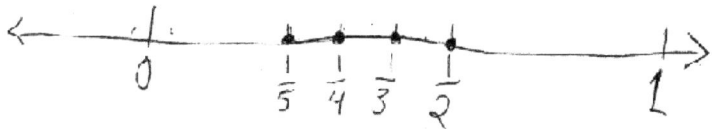

Figure 6. Abigail's reconstruction of the computer graph.

Considerations of distances between points representing rational numbers, not just their relative order, need to be brought to the attention of students. They need to use tools such as rulers to manually subdivide intervals into parts of varying size, and to obtain successive partitions. Rulers in the English system are convenient for subdivisions in 2, 4, and 8 parts. Metric rulers with mm subdivisions can be used for subdivisions into 2, 5, and 10 parts. Rulers that have an increasing number of subdivisions and that can be overlapped (Master Innovations 2005) can be used to help students see equivalence of fractions, and also focus on distances between fractions. Appropriate intervals can be constructed by the teacher or students so that students can evenly subdivide them with rulers into other numbers of parts.

Students in the upper elementary grades can also subdivide intervals by tracing parallel lines. This method allows them to subdivide a segment into an arbitrary number of parts. It will also lay the ground for important geometrical concepts such as equality of corresponding angles, and the relation between parallels in triangles and proportionality of segments generated by them. To divide a segment AB in, say, six parts, trace an auxiliary ray from B, and mark six equal intervals with a ruler (Figure 7). Connect C the endpoint of the sixth interval with A, and trace parallels to AC. The points of intersection of these line segments with segment AB will divide AB into six equal parts (Figure 8).

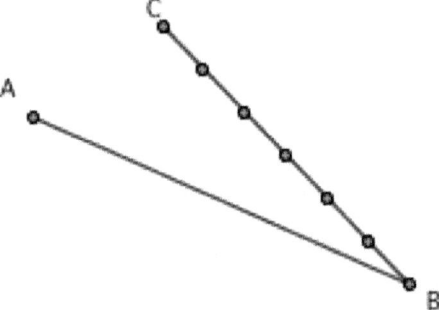

Figure 7. Equal intervals on auxiliary ray.

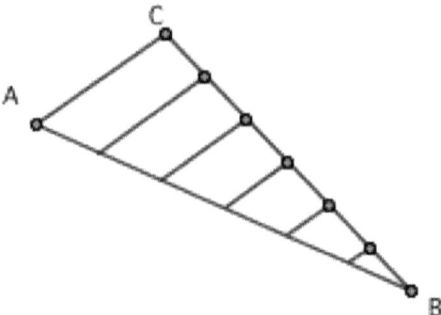

Figure 8. Parallel lines subdivide AB

Students can use a ruler and a construction triangle to trace the parallels. This will provide a kinesthetic approach to subdivision. The process is as follows. In the first step one edge of the construction triangle connects AC (Figure 9). The ruler is placed along the other leg of the construction triangle and is held firmly in place. In the second step (Figure 10), while holding the ruler firmly in place, the triangle is slid to the next mark on BC. Now hold the triangle firmly and trace the parallel line. Next, hold the ruler firmly again and slide the triangle to the next mark (Figure 11).

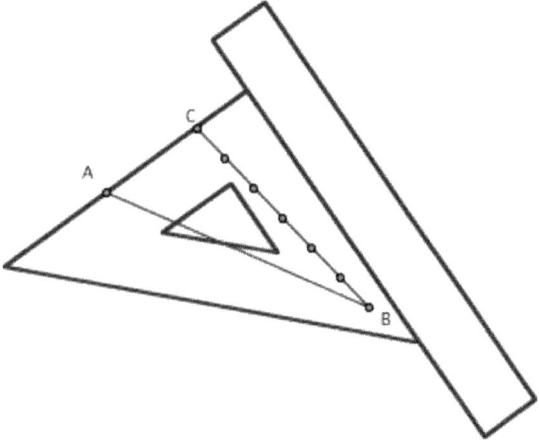

Figure 9. The edge of the construction triangle connects AC.

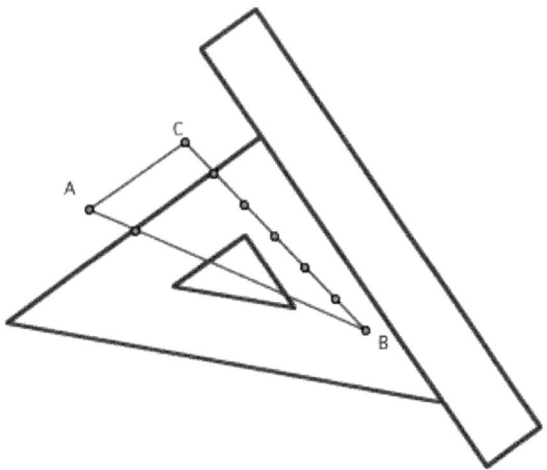

Figure 10. The triangle slides to the next mark on BC

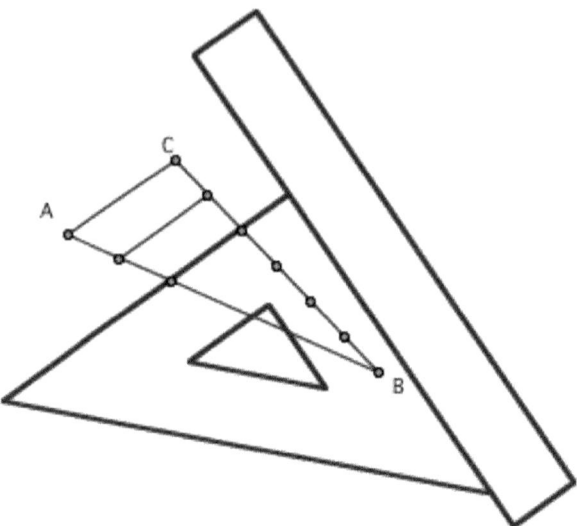

Figure 11. One more slide.

As students meet situations that require more accurate measures, they will begin working with the concepts of partitioning, and fraction equivalence. Students need to focus on successively partitioning the unit (Lamon, 2001). Students will see that the amount of partitions made determine the name of the fraction. To build a solid understanding, students need to work hands-on with measuring instruments over extended periods of time, while exploring real life situations.

An important goal in the teaching of rational numbers is to help students see a rational number not only as a relation between two numbers, but as a number in its own right, as a single object. Fraction equivalence is emphasized by the measure interpretation through the use of the number line. Intervals that are divided into different numbers of equal subdivisions and that can be compared to each other can help students see fractions that are equivalent. Because there is only one point associated with a given distance from zero, the fractions corresponding to the same point are equivalent (Figure 12).

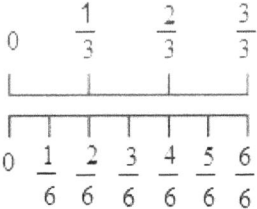

Figure 12. Equivalent fractions, 2/6 = 1/3, 4/6 = 2/3, ... ,

Students should locate fractions on a number line that is already partitioned for them, as well as lines without any partitioning already in place. Reverse problems are also important. Have students determine where 1 would be, given the location of 0 and a fraction already marked on the line, such as 1/2, 7/8 or 3/4. A good way to develop these skills is through hands-on practice with lots of problems that focus on partitioning. Accuracy is required in students' drawings to correctly develop understanding.

The number line is also useful to teach students about the density property of rational numbers. That is, students need to understand that between any two fractions there is another fraction (and therefore an infinite number of fractions), and that we can find a fraction as close as we like to any point. Students will not understand this just by hearing this; they need to experience these situations for themselves. They can engage in situations such as finding three fractions between 1/3 and 1/4. Teachers will know that they have provided a solid foundation of the measure interpretation when their students can: a) comfortably perform partitions other than halving, b) find any number of fractions between two fractions, c) use a given interval to measure any distance from the origin (Lamon, 2001).

Concluding remarks
Adding the quotient and measurement to the interpretations that students encounter with rational numbers will help them develop a richer and deeper understanding of rational numbers. Taken together with the part-whole model, they form a firmer foundation. As pointed out in the beginning, there are other contexts and interpretations for rational numbers that are important for student development, namely, ratio and operators (see Lamon 1999, ch. 7 and 11). Teachers need to realize that developing this multiple-faceted understanding of rational numbers takes several years and that it will not happen automatically. Students need to start developing these concepts in the elementary grades and continue during the middle grades. NCTM's *2002 Yearbook* is an excellent source of additional ideas (Litwiller 2002).

References

Arbaugh, Fran, Catherine Brown, Kathleen Lynch, and Rebecca McGraw. "Students' Ability to Construct Responses (1992-2000): Findings from Short and Extended Constructed-Response Items." In *Results and Interpretations of the 1990-2000 Mathematics Assessments of the National Assessment of Education Progress*, edited by Peter Kloosterman and Frank K. Lester, 337-362. Reston, Va.: National Council of Teachers of Mathematics, 2004.

Behr, Merlyn J., Guershon Harel, Thomas Post, and Richard Lesh. "Rational Number, ratio, and proportion." In *Handbook of research on mathematics teaching and learning*, edited by Douglas A. Grouws, 147-164. New York: Macmillan, 1992.

Cramer, Kathleen, Merlyn Behr, Richard Lesh, and Thomas Post. *The Rational Number Project Faction Lessons: Level 1.* Dubuque, Io.: Kendall/Hunt Publishing, 1997. Retrieved September 15, 2005 from http://education.umn.edu/rationalnumberproject/rnp1.html

Empson, Susan B. "Organizing Diversity in Early Fraction Thinking." In *Making Sense of Fractions, Ratios, and Proportions 2002 Yearbook*, edited by Bonnie H. Litwiller, 29-40. Reston, Va.: National Council of Teachers of Mathematics, 2002.

Flores, Alfinio, and Erika Klein. Connections between division and fractions in the third grade. *Teaching Children Mathematics, 11* (2005), pp. 452-457.

Lamon, Susan J. *Teaching Fractions and Ratios for Understanding*. Mahwah, N.J.: Lawrence Erlbaum Associates, 1999.

Lamon, Susan J. (2001). "Presenting and Representing from Fractions to Rational Numbers." In *The roles of representation in school mathematics 2001 Yearbook*, edited by Albert A. Cuoco, 146-165. Reston, Va.: National Council of Teachers of Mathematics, 2001.

Litwiller, Bonnie. *Making Sense of Fractions, Ratios, and Proportions, 2002 Yearbook*. Reston, Va.: National Council of Teachers of Mathematics, 2002.

Master Innovations. *The Master Ruler.* Available on line http://www.themasterruler.com/ Retrieved September 18, 2005.

Rothery, Thomas. *Abigail's conceptions of the number line.* Unpublished manuscript. Arizona State University, 2005.

Toluk, Zulbiye. Children's conceptualization of the quotient subconstruct of rational numbers. Unpublished doctoral dissertation, Arizona State University, 1999.

Wearne, Diana, and Vicky L. Kouba. "Rational Numbers." In *Results from the Seventh Mathematics Assessment of the National Assessment of Educational Progress*, edited by Edward A. Silver and Patricia Ann Kenney, 163-91. Reston, Va.: National Council of Teachers of Mathematics, 2000.

6 THE RULE OF FALSE POSITION AND GEOMETRIC PROBLEMS[6]

Introduction

Historical examples in college mathematics courses can help students understand the process of creation in mathematics and how mathematicians have grappled with problems over the ages. This can enliven mathematics and humanize it [1]. Swetz proposes to have students solve some of the problems that were of interest for early mathematicians as a direct approach to enrich mathematics teaching and learning through history [7]. Historical remarks can help students understand the material better and help them see how it fits into the wider domain of mathematics [5]. Historical examples of problems can also provide an interesting background to tell the history of mathematical ideas [4]. Future teachers can benefit especially from the historical perspective. They will learn solution methods alternative to the ones usually taught in schools. Let us look at a problem Simon Stevin published in 1583 [6].

First Problem: Construct a square knowing the difference PQ between its diagonal and its side.
Solution: Let BCDE be any square. On the diagonal EC take the point F so that EF = ED.

---

[6] Meavilla Seguí, V. and Flores, A. (2006). The rule of false position and geometric problems. *Convergence, 3.* On line journal
https://www.maa.org/book/export/html/116439. Used by permission of Mathematical Association of America.

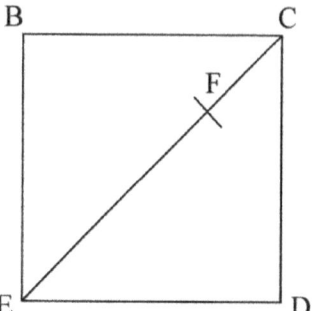

Figure 1. Construction of square

If $FC = PQ$, then the square $BCDE$ is the solution to the problem. If not, the side of the solution square (say y) will be the fourth proportional with respect to the segments FC, PQ and ED. That is, $\dfrac{FC}{PQ} = \dfrac{ED}{y}$

Simon Stevin

Stevin was born in Bruges, Flanders (now Belgium) in 1548 and died at The Hague (Netherlands) in 1620. He introduced the systematic use of decimal numbers into European mathematics, and proposed the unification of the system of weights and measures with a method based on the decimal subdivision of the unit. He also published one of

The method used by Stevin is called the *rule of false position*. The solution is especially interesting because he used this method in a geometrical problem; the method was mainly used in problems of algebraic nature. It is also interesting because our students do not use geometrical proportionality when they face problems about constructing figures that have some restrictions. To solve the problem of the square and its diagonal, students in high school (ages 16-18) frequently used methods involving trigonometry and algebra to compute the side of the square. For them the real challenge is to construct the figure.

Construction of the fourth proportional with respect to three segments

Let a, b, and **c** be three line segments.

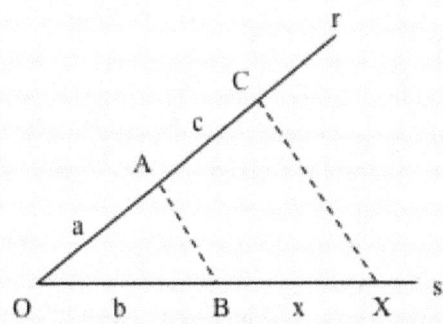

On ray Or trace segments OA = a and AC = c. On ray Os trace segment OB = b. From point C trace CX parallel to AB. We can verify that x is the fourth proportional with respect to the segments a, b, and c. That is,

$$\frac{a}{b} = \frac{c}{x}$$

Indeed, triangles OAB and OCX are similar. Therefore

$$\frac{a}{b} = \frac{a+c}{b+x} \Rightarrow \frac{a}{b} = \frac{(a+c)-a}{(b+x)-b} \Rightarrow \frac{a}{b} = \frac{c}{x}$$

The rule of false position

The *rule of false position, regula falsi, rule of one false position, or simple rule of false position* was very popular in mathematics texts of the 16th century, and was still present in some books of elementary mathematics in the fist half of the 20th century [4]. Lumpkin [2] traces the development of the method of false position from its uses in ancient Egypt, the Hellenistic world, medieval Islam, to its transmission to Europe and subsequent appearance in mathematical texts in the eighteenth century, to Benjamin Bannaker's advancement of the rule of false positions through his own efforts.

In general, *regula falsi* was used to solve equations of first degree with one unknown, without using algebraic symbolic notation. The statements of problems that were solved using the rule of false position can be translated to an equation of the type $ax = b$ or more precisely, $a_1x + a_2x + \ldots + a_nx = b$.

We describe the simple rule of false position with modern algebraic notation:

Suppose we want to solve the equation $ax = b$ (1).

If we make $x = c$, then $ac = b_1$ (2).

There are two possibilities:

a) If $b_1 = b$, then $x = c$ is the solution of the equation.

b) If $b_1 \neq b$, dividing term by term the equations (1) and (2) it results that: $\dfrac{x}{c} = \dfrac{b_1}{b}$

Geometric justification of the rule of one false position

Regula falsi can be justified geometrically (see Figure 2). Using similar triangles we have:

$$\frac{c}{b_1} = \frac{x}{b} \quad \Rightarrow x = \frac{bc}{b_1}$$

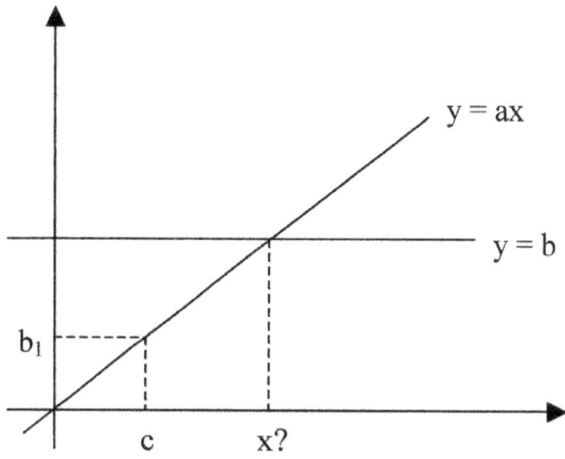

Figure 2. Geometric justification of regula falsi

Regula falsi and geometric problems

The mathematician, physicist, engineer, and inventor Simon Stevin used the rule of one false position in the solution of additional problems of geometric nature. He included his geometric investigations in the *Problematum geometricorum* (1583), which he structured in five books (Figure 3).

In the second book (De continuae quantitatis regula falsi) Stevin used the rule of one false position in the solution of four problems, doing some geometrical constructions with the help of similar figures. In the solution of each of the problems there are five phases: (1) Display of the data, (2) Display of the problem, (3) Construction, (4) Demonstration, (5) Conclusion. In the third phase the simple rule of false position is used. Stevin followed a plan similar to the one described in Figure 4.

PROBLEMATVM

GEOMETRICORVM

In gratiam D. MAXIMILIANI, DOMINI A CRVNINGEN &c. editorum, Libri v.

Auctore

SIMONE STEVINIO BRVGENSE.

ANTVERPIAE,
Apud Ioannem Bellerum ad infigne
Aquilæ aureæ.

Figure 3. Stevin's Problematum Geometricorum

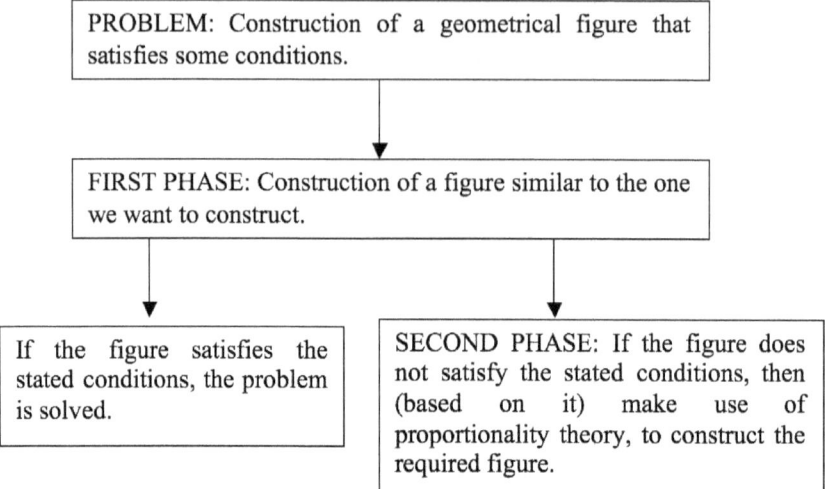

Figure 4. Phases to solve geometrical problems with regula falsi

We will present problems 1, 3, and 4 (problem 2 was presented at the beginning of this article). Given the expository nature of this article, we adapted the problems from the original text to modern language. We have tried to adhere to the spirit of the author.

Second problem

Build an equilateral triangle knowing the length of a line segment *PQ* equal to the side of the triangle less its height plus a third of its height.

Construction

Let *BCD* be any equilateral triangle. Draw the height *BE* and segment *FG* that joins the center of the triangle with the midpoint of the side *BC*. On the side *CD* take the point *H* so that *CH* = *BE*. On the extension of *CD* take the point *I* so that *DI* = *FG*.

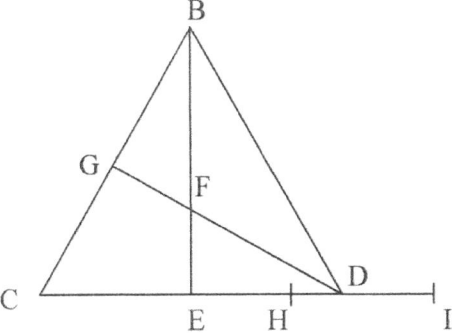

Figure 5. Construction of equilateral triangle

If triangle BCD is the solution to the problem, then HI will be equal to the line segment PQ. If not, the side of the equilateral triangle solution (say x) will be the fourth proportional with respect to the segments HI, PQ and BC. In other words: $\dfrac{HI}{PQ} = \dfrac{BC}{x}$.

Third problem

Build a regular pentagon knowing one segment PQ whose endpoints are one of the vertices of the pentagon and the midpoint of the opposite side.

Construction

Let $BCDEF$ be any regular pentagon. Draw the line segment BG that joins the vertex B with the midpoint of the side ED.

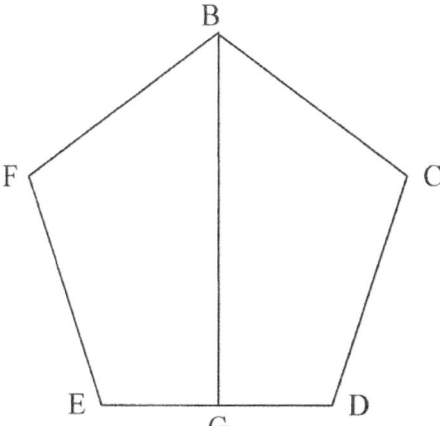

Figure 6. Construction of pentagon

If $BG = PQ$, then the regular pentagon $BCDEF$ is the solution to the problem. If not, the side of the pentagon solution (say z) will be the fourth proportional of the segments BG, PQ and ED. That is, $\dfrac{BG}{PQ} = \dfrac{ED}{z}$.

Fourth problem

Let $RSTUV$ be a given polygon and PQ a given line segment. Construct a polygon $MNKIL$ similar to the previous and equally arranged so that if segment MN, homologous to RS, is taken from LN, homologous to VS, and to the rest you add segment LI, homologous to VU, you obtain a segment equal to PQ.

Construction

Let $BCDEF$ be any polygon similar to $RSTUV$ and equally arranged. On segment CF take point G so that $CG = CB$. Extend segment CF to point H so that $FH = FE$.

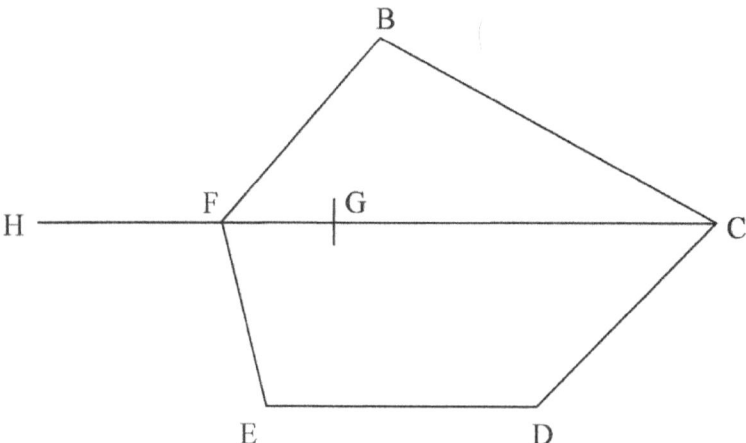

Figure 7. Construction of polygon

If $HG = PQ$, then the polygon BCDEF is the solution to the problem. If not, we will proceed in the following way:
We will determine the fourth proportional (say w) with respect to the segments HG, PQ and ED. Then, $w = IK$ will be the segment homologous to ED in the solution polygon. From it the required polygon will be constructed.

Conclusion

When students solve problems similar to the ones presented, the following contents of conceptual, procedural, or attitudinal nature can be introduced, firmed, or developed.

Conceptual contents
- Proportional magnitudes.
- Congruence and similarity of figures. Similarity ratio.
- Fourth proportional with respect to three line segments.

Procedural contents
- Identify the measure characteristics of similarity of plane figures in concrete situations.
- Apply similarity of figures to the solution of construction problems.
- Construction of the fourth proportional with respect to three line segments.
- Identification of similarity in plane figures.
- Search for relations in geometric figures and configurations.
- Construction of plane figures.

Attitudinal contents
- Value the usefulness of proportionality for the solution of problems.
- Skillful utilization of some drawing instruments (ruler, construction triangle, compass, computer programs…).
- Recognize and value the usefulness of regula falsi for the solution of geometrical problems.
- Flexibility to solve geometric problems with different approaches.

In addition to the above benefits, the use of regula falsi can encourage teachers and students to experiment in mathematics. Experimenting in the context of similarity problems can lead to a better understanding. Discovery in mathematics is frequently preceded by multiple experiments.

References

1. AVITAL, S. (1995). History of mathematics can help improve instruction and learning. In F. J. Swetz, J. Fauvel, O. Bekken, B. Johansson & V. Katz (Eds.), *Learn from the masters!* (pp. 3-12). Washington, DC: Mathematical Association of America.

2. LUMPKIN, B. (1996). From Egypt to Benjamin Bannaker: African origins of false position solutions. In R. Calinger (Ed.), *Vita mathematica* (pp. 279-289). Washington, DC: Mathematical Association of America.

3. MEAVILLA, V. (2000). Historia de las Matemáticas: métodos no algebraicos para la resolución de problemas. *SUMA. Revista sobre la enseñanza y el aprendizaje de las Matemáticas*, n° 34, pp. 81-85.

4. MEAVILLA, V (2001). *Aspectos históricos de las matemáticas elementales.* Zaragoza, Spain: Prensas Universitarias de Zaragoza.

5. RICKEY, V. F. (1996). The necessity of history in teaching mathematics. In R. Calinger (Ed.), *Vita mathematica* (pp. 251-256). Washington, DC: Mathematical Association of America.

6. STEVIN, S. (1583). *Problematum geometricorum.* Antverpiae, Apud Ioannem Bellerum.

7. SWETZ, F. J. (1995). Using problems from the history of mathematics in classroom instruction. In F. J. Swetz, J. Fauvel, O. Bekken, B. Johansson & V. Katz (Eds.), *Learn from the masters!* (pp. 25-38). Washington, DC: Mathematical Association of America.

8. WINICKI, G. (2000). The analysis of regula falsi as an instance for professional development of elementary school teachers. In V. Katz (Ed.), *Using history to teach mathematics: An international perspective* (pp. 129-133). Washington, DC: Mathematical Association of America.

On-line references

1. Biographies of Simon Stevin
http://www-groups.dcs.st-and.ac.uk/~history/Mathematicians/Stevin.html
http://www.divulgamat.net/weborriak/Historia/MateOspetsuak/Stevin.asp (in Spanish)
http://www.bbc.co.uk/history/historic_figures/stevin_simon.shtml

2. Works of Simon Stevin
http://www.library.tudelft.nl/ws/729/f_NL.html.

7 POSING PROBLEMS TO DEVELOP UNDERSTANDING: TWO TEACHERS MAKE SENSE OF DIVISION OF FRACTIONS[7]

In this article we discuss first the benefits of students posing their own problems in mathematics. Then we discuss briefly the importance of teachers posing problems for their students, and finally we describe in detail how two teachers posed plenty of problems to develop their own understanding of a mathematical topic.

Students pose their own mathematical problems
Learning to pose problems is important not only in school mathematics. Kilpatrick (1987) points out that in real life many problems need to be first posed by the solver, who initially formulates the problem. Students who pose problems in mathematics can become better at solving problems. For instance, one of the heuristic strategies suggested by Pólya (1957) to deal with difficult problems is to pose other problems related to the problem but that are easier to solve. Another strategy is to modify the problem, for example, by keeping only part of the condition. Brown and Walter (1993, p. xiv) point out that one can create "deeper insights about anything that is *given* by modifying the given." The *Professional standards for teaching mathematics* (National Council of Teachers of Mathematics 1991) state that "Students should be given opportunities to formulate problems from given situations and create

---

[7] Flores, A., Turner, E. E., and Bachman, R. C. (2005). Posing problems to develop understanding: Two teachers make sense of division of fractions. *Teaching Children Mathematics*, *12*, 117-121. Copyright National Council of Teachers of Mathematics. Used by permission.

new problems by modifying the conditions of a given problem." (p. 95). Silver (1994) highlights the importance of students posing mathematical problems through different perspectives. These include problem posing as

- a feature of inquiry-oriented instruction;
- a prominent feature of mathematical activity;
- a means to improve students' problem solving;
- a window into students' mathematical understanding; and
- a means of improving student disposition toward mathematics.

Students who pose their own problems are more motivated to solve them than if the problems come from an outside source such as the textbook. Silver and Cai (1996) found that students are quite able to generate interesting problems within arithmetic contexts. In a study with 509 middle school students they found that students are able to generate a large number of problems that are mathematically solvable. Many of those problems are complex both in their form and their content. Nearly half the students generated sets of related problems. Silver and Cai also found that students who are more successful in solving problems generate more mathematical problems and more complex problems.

We need also be aware that the context in which the problems are posed is important. English (1998) working with third graders (8 years old) found that children generate a broader range of problem types for number situations that are informal and without a formal symbolism. English (1997) describes the use of questions that lead to further ideas and questions with some multicultural activities to develop a problem-posing classroom with fourth and fifth graders.

Students who pose problems develop a better understanding of the mathematics involved. As Brown and Walter (1992) state, coming to know something requires that "we operate or even modify the things we are trying to understand" (p. 2). Posing problems is a great way to do that. Moses, Bjork, and Goldenberg (1993) point out that "we learn mathematics particularly well when we are actively engaged in creating not only the solution strategies but the problems that demand them" (p. 187).

Teachers pose mathematical problems for their students
In addition to letting students pose their own problems, teachers need to develop the ability to pose problems for their own students. Using the textbook and curricular guides, the teacher can lay out the map of the concepts and skills students need to learn. Well-chosen problems can stimulate mathematics learning. Some of the problems will be adapted from

textbooks and other curriculum materials. Some will be the teachers own problems. As *Principles and Standards* point out,

> The teacher's role in choosing worthwhile problems and mathematical tasks is crucial. By analyzing and adapting a problem, anticipating the mathematical ideas that can be brought out by working on the problem, and anticipating students' questions, teachers can decide if particular problems will help to further their mathematical goals for the class. (National Council of Teachers of Mathematics 2000, p. 52)

Teachers can pose problems for their students to make sense of division of fractions. Children (ages 11-12), who work on their own on a sequence of problems posed by the teacher, and use pictorial representations in terms of measurement interpretation of division of fractions, can observe patterns of the numbers involved (Pirie, 1988). By looking at the patterns, some children will discover that they can solve a division problem by multiplying the first fraction by the denominator of the second fraction, and then divide by the numerator of the second. As one girl expressed it, "You just times the bottom number and see how many 2's or 4's or whatever go into it, like if it was divided by 4/5 then times by 5 and see how many 4's or whatever go into it" (Pirie, p. 3). The same girl, a few weeks later, expressed the division process of fractions more concisely as "You just turn it upside down and multiply" (Pirie, p. 4). However, it is unlikely that students without the proper guidance from the teacher will be able to take the next step, that is, to explain why "invert and multiply" works, especially in terms of the role of reciprocals and the inverse relation between multiplication and division. Teachers need to be able develop their own understanding to provide this guidance.

Teachers pose mathematical problems for themselves

There is another reason why posing problems is also important for teachers: the posing of mathematical problems can help teachers develop their own understanding. Teachers who for the most part went through their own schooling learning "rules without reasons" in mathematics, will have to learn how to teach for conceptual understanding. It is important to document and understand how teachers can learn to make sense of the topics they teach, and furthermore, how they can learn to make sense also in terms that are meaningful for their own students. Learning to make sense is by no means an easy task. For example, Lubinski, Fox, and Thomason (1998) documented in detail that it is possible for a prospective elementary teacher to develop understanding of division of fractions on her own. At the same time, the study showed how difficult it was for the prospective teacher to gain a basic conceptual understanding of division of fractions. In that study, the prospective teacher was asked as an assignment that extended over three weeks to develop her own understanding of division of fractions. In one of her attempts to make sense, the future teacher used more advanced mathematical tools, such as an algebraic method, to gain procedural understanding of why you invert and multiply when dividing fractions. She was not satisfied with that explanation for division of fractions, and she struggled with different approaches until she was finally able to understand conceptually using the measurement interpretation of division. However, as Lubinski et al. point out, the prospective teacher was not ready yet to think about the central role of reciprocals and the inverse relationship between multiplication and division, a crucial aspect if one wants to have a deeper conceptual understanding of the invert and multiply algorithm.

In this section we will describe how two teachers used problem posing very effectively to develop their own conceptual understanding of division of fractions in terms that would also be helpful for their students. Two teachers, Elizabeth and Carolyn, participated in this study. Each one was teaching a combined fourth-fifth grade in an urban elementary school. Carolyn has a master's degree in education and at the time had eleven years of teaching experience. Elizabeth was in the process of getting a master's degree with emphasis in mathematics education and was in her fourth year of teaching. Both teachers were recognized in their own school and district as expert teachers and were frequently asked to help prepare other teachers in areas such as language arts, science, and technology. The two teachers had a double role in the study, as the subjects of study and also as active participants deciding what aspects were relevant. Examples of the problems posed and solved by the teachers were collected during an initial session and from the draft of an article written by the teachers. Additional insights and information were obtained during later interviews.

In the first part of the study, during a three-hour session in a graduate methods course, the two teachers were given a kit of rectangular translucent colored fractions (*Dr. Loyd's Fraction Kit*, 1987). Both teachers were familiar with the kit as they had used it before in a methods course to illustrate the measurement interpretation of division of fractions. They were asked to collaborate to develop a lesson to teach division of fractions, including the connection to the algorithm of "invert the second fraction and multiply" in terms that would be meaningful to students in the upper elementary and middle grades. The two teachers worked independently from the instructor, except for brief interactions, where he pointed to possible routes such as working with a more accessible *special situation* (Pólya, 1962), such as using fractions with the same denominators. The teachers used the fraction kit heavily, posed and solved lots of problems, wrote some comments, and outlined a lesson for students in their notes. They then presented the main ideas of their lesson to the other teachers in the course at the end of the session.

Both teachers kept a very detailed written record of their efforts. Elizabeth and Carolyn posed a great variety of problems, starting with the ones that facilitate understanding of the situation rather than obscure it. Elizabeth posed 64 of her own problems in 15 handwritten pages, and Carolyn posed 45 problems of her own in 11 pages. This detailed record made it possible to reconstruct their experience. Their notes show that they were able to make the connection between their previous understanding of division of fractions and the standard algorithm in that session.

Elizabeth's and Carolyn's approach when learning a new topic or idea is to first "play." They started by posing and solving simple problems, such as $\frac{1}{2} \div \frac{1}{2}, \frac{1}{2} \div \frac{1}{4}, \frac{1}{2} \div \frac{1}{6}$, and $\frac{1}{2} \times 6$. They used the concrete fraction models, and also their procedural knowledge of "invert the second fraction and multiply." Their notes reveal that they worked in parallel, following the same categories of examples, but each one posing her own problems. They used a variety of problems, including some with whole numbers, some with fractions with same denominator, and some with fractions with different denominators, but where the answer was easy to see.

After about 20 problems, Carolyn worked with the case $\frac{2}{3} \div \frac{1}{4}$. Her notes show that she changed it to $\frac{8}{12} \div \frac{3}{12}$, wrote then $8 \div 3$ and then $2\frac{2}{3}$. Her next entry reveals that she has noticed something important, that when dividing fractions with the same denominator, the denominator will not appear as part of the answer:

$$\frac{15}{\cancel{21}} \div \frac{7}{\cancel{21}}$$

She then wrote the answer as $\frac{15}{7}$, and as mixed number.

In the meantime, Elizabeth was working in parallel. After about twenty examples, she wrote the following observation with the two examples

$$\frac{5}{6} \div \frac{2}{6} = 2\frac{1}{2} \qquad \frac{5}{8} \div \frac{2}{8} = 2\frac{1}{2}$$

"When we are working with pieces of the same size, what matters is the numerator (the number of pieces)." Using the measurement interpretation she had realized that dividing 5 pieces by 2 pieces of the same size would give the same result. It did matter the size of the pieces being compared in the division problem as long as the pieces representing the dividend and divisor were the same size. Both teachers wrote their answers most of the times as mixed numbers (rather than improper fractions), a notation that is closer to the meaning of "how many times does it fit", but obscures the relation between the numerators of the fractions in the division problem. Then each one posed and solved about 12 more problems, with fractions with different denominators, converting them to common denominator, and dividing numerators afterwards.

At this point, their attention shifted to the way common denominators are obtained and its connection to multiplication. For the problem $\frac{2}{5} \div \frac{1}{3}$, Elizabeth's notes show that she finds equivalent fractions with common denominators, and she highlights the process

$$\begin{array}{cc} & \times\,5 \quad \frac{5}{} \\ \times\,3 \quad \frac{6}{} & \times\,5 \quad \frac{5}{15} \\ \times\,3 \quad \overline{15} & \end{array}$$

Several examples later there is a revealing entry in her notes

$$\frac{2\times 5}{3} \underset{}{\overset{}{\times}} \frac{4}{5} \,\,\rangle$$

The arrows indicate the place where numerator and denominator of the second fraction will have an effect.

They posed and solved several more problems. The last example in Carolyn's notes (before the lesson plan) shows that the connection has been made between their previous understanding division of fractions in terms of measurement and the standard rule of multiplying by the inverse, via fractions with the same denominators. She labels the example "Rule to divide"

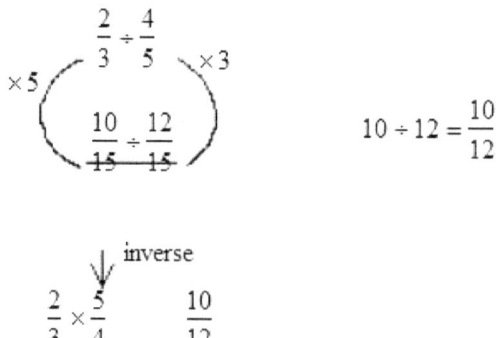

After this extensive "play" of posing and solving problems they finally felt ready to start planning the lesson. They knew exactly what kinds of problems to pose. They would also start by dividing fractions with the same denominators. Elizabeth wrote "We are going to give them many examples so that they can see that the denominator does not matter — and we are going to present examples where the numerator does not change so that they see the pattern."

The notes of the teachers reveal that also during this planning each one posed her own problems, but following a common plan. They posed problems with simple and familiar fractions and moved gradually to more general cases. The first five steps described below are from Elizabeth's notes, in addition, Carolyn included a sixth step.

1) Use fractions with the same denominator, so that the first is bigger than the second, and the quotient a whole number, for example $\frac{4}{6} \div \frac{2}{6}$.

2) With the same conditions, but the quotient does not have to be a whole number, $\frac{5}{6} \div \frac{2}{6}$.

3) Use fractions that do not have the same denominator, but that are well known, and that are related, that have common factors, such as, $\frac{1}{2} \div \frac{1}{4}$, and $\frac{1}{6} \div \frac{1}{3}$. With these examples, we want that the children can find, by using manipulatives, the common denominator. What smaller piece will fit into the two fractions?

4) Now with more difficult examples, as with $\frac{2}{3} \div \frac{2}{4}$. They also need to be able to find common denominator, $\frac{8}{12} \div \frac{3}{12}$.

5) Ask students how we obtain the answer, look at the numbers, and search for patterns and similarities $\frac{2 \times 4}{3 \times 4}$ $\frac{8}{12}$. They are multiples/factors.

6) Look at the original problem $\frac{2}{3} \div \frac{1}{4}$ (multiplying by 4) [Carolyn's emphasis]. Explain why the algorithm works.

After the plan was laid out, the last set of problems brings them to the desired connection. One entry in Elizabeth's notes shows

$$\frac{(2 \times 4)}{3 \times 4} \div \frac{(2 \times 3)}{4 \times 3}$$

She wrote the problem later as $\frac{8}{12} \div \frac{6}{12} = \frac{8}{6}$. Notice that the answer this time was written as an improper fraction, a notation that brings to the forefront the relation between numerators. Elizabeth made a big note about the relation between fractions and quotients of whole numbers

$$\frac{2}{3} = 2 \div 3 \qquad \frac{3}{5} = 3 \div 5$$

Elizabeth and Carolyn presented their approach to the rest of the class. During their presentation, they outlined the parts described above and summarized some of the key points.

When dividing fractions with the same denominator the result is simply the fraction that corresponds to the quotient of the numerators, the denominator does not play a role in the answer. In the case of fractions with unequal denominators, we need to obtain common denominators first. To accomplish this, we multiply the numerator and denominator of the first fraction by the denominator of the second, and we multiply the numerator and denominator of the second fraction by the denominator of the first. To obtain the final answer, we need to pay attention only to the new numerators, and divide them.

The following diagram, found in Elizabeth's notes, encapsulates how the process of obtaining common denominators will give the two terms of the final answer.

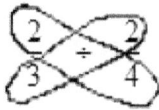

When converting to common denominators, the numerator of the first fraction will be 2×4, the numerator of the second 3×2. The answer when dividing the two new fractions will be $\frac{2 \times 4}{3 \times 2}$, the same result as if we had multiplied $\frac{2}{3} \times \frac{4}{2}$. Notice that the fraction $\frac{4}{2}$ is the inverse of $\frac{2}{4}$.

Three months after the two teachers finished their lesson, the instructor suggested to the teachers to write a paper for a journal for teachers of mathematics to bridge a common gap in the preparation offered to teachers, namely between the concrete hands-on approach to division of fractions and the symbolic algorithm of "invert and multiply."

The two teachers focused mainly on what they had done in their lesson, together with describing necessary experiences and knowledge that students would need to be able to benefit from it. They also included problems in contexts such as baking, distance, fabric and sewing, food, and money. For example, "You need 2 cups of flour for a recipe, you have a 1/2 cup measuring utensil. How many 1/2 cups are in 2 cups?" This is how Elizabeth and Carolyn described the relation between the method of using common denominator and the algorithm to divide fractions:

> To help them understand why this algorithm works we would return [to] our previous ways of solving the problem through the finding of common denominators. $\frac{2}{3} \div \frac{1}{4} = \frac{8}{3}$. We will ask the students, What did you do when you converted $\frac{2}{3}$ into $\frac{8}{12}$? They would talk about multiplying the numerator and denominator by 4. We will also ask how they

converted $\frac{1}{4}$ into $\frac{3}{12}$? This time they will talk about multiplying the numerator and denominator by 3. So we would discuss the relationships between finding common denominators and using the algorithm. When they found a common denominator for $\frac{2}{3}$ they multiplied 2×4 to get a numerator of 8. In the algorithm they are also multiplying 2×4 to get 8 in the numerator. When they found the common denominator for $\frac{1}{4}$ they multiplied 1×3 to get a numerator of 3. Then in the algorithm they are also multiplying 1×3 to get 3 in the denominator of the answer. (The reason they multiply by 4 and 3 in the algorithm is to create fractions with common denominators without having to go through multiple steps).

The two teachers had bridged together the concrete approach with the algorithm of multiplying by the inverse, but so far they had not brought to the front the role of reciprocals and the inverse relationship between multiplication and division. However, as the teachers were writing the last paragraph of their first draft, they had an insight that had not surfaced during the lesson. The two teachers made the following statement in the last line: "The reason they invert the second fraction is because of the inverse relationship between multiplying and dividing." The instructor pointed that they had just hit the core of the issue, and encouraged them to elaborate on this connection and include it in their paper. The connections established by the teachers during this subsequent phase are described below.

To help students realize the inverse relationship between multiplication and division, Elizabeth and Carolyn think that it is important for children to go back and look at examples of multiplication of whole numbers, so that students could really consider the inverse relationship. They suggested posing problems such as:

$8 \div 2 = 4, \qquad 4 \times 2 = 8, \qquad 8 \div 2 \times 2 = ?$

$6 \div 2 = 3, 3 \times 2 = 6, 6 \div 2 \times 2 = ?$

and posing the reverse problems, that is where an operation would "undo" the previous operation and give back the original number

$8 \times 4 = 32, \qquad 32 \div 4 = 8, \qquad 8 \times 4 \div 4 = ?$

$10 \times 2 = 20, \qquad 20 \div 2 = 10, \qquad 10 \times 2 \div 2 = ?$

They would discuss these problems with students and help them notice that multiplication and division are inverse operations. They would help them make a connection to other operations, such as the inverse nature of addition and subtraction. Informal language used by students such as saying that the two operations "undo" each other, or "cancel" each other, would be accepted, but the teachers would also use the vocabulary of inverse operations.

Elizabeth and Carolyn would then introduce examples that involve fractions, such as

$$\frac{3}{4} \div \frac{1}{4} = 3, \quad 3 \times \frac{1}{4} = \frac{3}{4}, \quad \frac{3}{4} \div \frac{1}{4} \times \frac{1}{4} = ?$$

$$\frac{4}{6} \div \frac{2}{6} = 2, \quad 2 \times \frac{2}{6} = \frac{4}{6}, \quad \frac{4}{6} \div \frac{2}{6} \times \frac{2}{6} = ?$$

We can notice that from the equation $\frac{4}{6} \div \frac{2}{6} \times \frac{2}{6} = \frac{4}{6}$, there is just two more steps to the "invert and multiply" algorithm. If we multiply both sides of the equation by $\frac{6}{2}$, we get $\frac{4}{6} \div \frac{2}{6} \times \frac{2}{6} \times \frac{6}{2} = \frac{4}{6} \times \frac{6}{2}$, and simplifying the product of reciprocal fractions, we obtain $\frac{4}{6} \div \frac{2}{6} = \frac{4}{6} \times \frac{6}{2}$. This was, however, not the route that Elizabeth and Carolyn followed, although apparently it was well within their reach.

They had used variables and properties of equations such as "multiplying both sides by the same number preserves the equality", when they themselves were working "backwards" trying to make sense of the algorithm that had learn years before. However, they preferred not to use this approach of "working backwards", of first giving the algorithm and then try to make sense of it, because they wanted their students discover the formulas or the algorithms. According to them, when children are given the algorithm, they lose the motivation to figure things out and to make sense. It is interesting to note that the avenues of exploration that these two teachers decided to follow in more depth were greatly influenced by their perception of what would be meaningful to their own students.

Another reason why Elizabeth and Carolyn did not take the last step to connect $\frac{4}{6} \div \frac{2}{6} \times \frac{2}{6} = \frac{4}{6}$ with $\frac{4}{6} \div \frac{2}{6} = \frac{4}{6} \times \frac{6}{2}$ was that although both teachers were quite comfortable solving for x in an equation like $x \cdot \frac{2}{6} = \frac{4}{6}$, they were not used to deal with $\frac{4}{6} \div \frac{2}{6}$ as a single mathematical object in an equation like $\frac{4}{6} \div \frac{2}{6} \times \frac{2}{6} = \frac{4}{6}$. They saw the left side of this equation as a concatenation of operations.

Elizabeth and Carolyn approach was to pose problems that illustrate that when you divide by a number, it is the same as multiplying by the inverse, so that students could see the connection. Again, they started with whole numbers. They showed that $4 \div 2$ is exactly the same as $4 \times \frac{1}{2} = 2$. To do this, they used the connection between fractions and division, $4 \div 2 = 2$, which is exactly like $\frac{4}{2} = 2$, which is equal to $\frac{4 \times 1}{1 \times 2}$ or $4 \times \frac{1}{2}$. Then they used examples with fractions, starting with easy fractions. However, to show that $\frac{1}{2} \div \frac{1}{4}$ is the same as $\frac{1}{2} \times \frac{4}{1}$, they used their first approach of using common denominators, rather than using the inverse relationship of multiplication and division.

Final remarks

We have seen how posing problems played a major role in the approach of these two teachers to make sense of division of fractions for themselves and their students. Quite remarkable was the great number of problems posed and solved by Carolyn and Elizabeth as they tried to develop understanding. Sometimes teachers pose only one problem when they or their students are trying make sense of a concept. Sometimes the problem posed is not the most revealing. Imagine how difficult it would be for a teacher or her students to make sense of division of fractions if they were to deal only with one problem like $\frac{2}{3} \div \frac{5}{7}$. The approach of Carolyn and Elizabeth was to pose several problems of various degrees of difficulty and complexity for each aspect of the situation. Their approach had also a cyclical nature of gradually incorporating more difficult problems for each aspect and returning again and again to pose easy to see and easy to deal problems when shifting to a new aspect. These features of posing a multiplicity of problems, of different degrees of difficulty, and illuminating different aspects of the situation, are also helpful when teachers pose problems for their own students, or when students pose their own problems.

References

Brown, Stephen I., and Marion I. Walter. *The Art of Problem Posing.* 2nd ed. Hillsdale, NJ: Lawrence Erlbaum Associates, 1990.

Dr. Loyd's Fraction Kit. St. Louis, MO: Pegasus Publications, 1987.

English, Lyn D. "Promoting a problem-posing classroom." *Teaching Children Mathematics,* 4 (1997, November): 172-179.

English, Lyn D. "Children's Problem Posing within Formal and Informal Contexts." *Journal for Research in Mathematics Education* 29 (1) (1998): 83-106.

Kilpatrick, Jeremy. "Problem Formulating: Where Do Good Problems Come From?" In *Cognitive Science and Mathematics Education,* edited by Alan H. Schoenfeld, 123-47. Hillsdale, NJ: Lawrence Erlbaum, 1987.

Lubinski, Cheryl A., Thomas Fox, and Rebecca Thomason. "Learning to Make Sense of Division of Fractions: One K-8 Preservice Teacher Perspective." *School Science and Mathematics* 98 (1998): 247-59.

Moses, Barbara N., Bjork, Elizabeth, and Goldenberg, E. Paul. "Beyond problem solving: Problem posing." In *Problem posing: Reflections and applications*, edited by Stephen I. Brown and Marion I. Walter, 178-188. Hilldale, NJ: Lawrence Erlbaum, 1993.

National Council of Teachers of Mathematics. *Principles and Standards for School Mathematics*. Reston, VA: National Council of Teachers of Mathematics, 2000.

National Council of Teachers of Mathematics. *Professional Standards for Teaching Mathematics*. Reston, VA: National Council of Teachers of Mathematics, 1991.

Pirie, Susan E. B. "Understanding: instrumental, relational, intuitive, constructed, formalised...? How can we know?" *For the Learning of Mathematics* 8 (1988): 2-6.

Pólya, George. *How to Solve It: A New Aspect of Mathematical Method*. 2nd ed. Princeton, NJ: Princeton University Press, 1957.

Pólya, George. *Mathematical Discovery: On Understanding, Learning, and Teaching Problem Solving*. Vol. 1. New York: John Wiley & Sons, 1962.

Silver, Edward A. "On Mathematical Problem Posing." *For the Learning of Mathematics* 14(1) (1994): 19-28.

Silver, Edward A, and Jinfa Cai. "An Analysis of Arithmetic Problem Posing by Middle School Students." *Journal for Research in Mathematics Education* 27 (5) (1996): 521-39.

8 CONNECTIONS BETWEEN DIVISION AND FRACTIONS IN THE THIRD GRADE[8]

Abstract:
This article describes ways in which third graders solved the problem of sharing 7 brownies among 4 people using concrete materials and the connections in fractions they can make with the guidance of the teacher.

Many elementary teachers find it challenging to help their students understand mathematical concepts and make connections to other concepts. The concepts related to fractions can be very difficult for some students to grasp. To give some insight as to how students think about division and fractions, this article presents strategies used by children to solve a fraction problem. It also illustrates how teachers can use these strategies to help students establish connections among different concepts related to fractions. Ms. Klein, a third grade teacher, posed the following problem to her students: Tonight my mom, dad, grandma, and myself will sit down to dinner. There will be seven brownies for dessert. How can we share the brownies so that everyone has the same amount? I really want to make sure my dad, who can be very sneaky when it comes to brownies, does not get a bigger share. (Adapted from Tierney and Berle-Carman, 1995.)

Before posing this problem to her class, Ms. Klein provided each student with a big sheet of paper with seven colored rectangles of the same size. The rectangles represented the seven brownies the students had to share. The

[8] Flores, A. and Klein, E. (2005). Connections between division and fractions in the third grade. *Teaching Children Mathematics*, *11*(9), 452-457. Copyright National Council of Teachers of Mathematics. Used by permission.

students were also given scissors, glue, and a piece of paper to display the processes they used to share the fractions and their solutions. Students worked on the problem on their own, often sharing some of their ideas with their partner as they developed their solution. As part of their solution strategy, many students displayed four people on the blank sheet of paper or divided it into four sections. Students started to put brownies and parts of brownies in the corresponding section, making sure each one received the same amount.

Students used different strategies to solve the brownie problem. Alexis gave one brownie to each person. Then she cut one of the remaining brownies in half, gave one piece to two of the people. She then cut another brownie in two equal parts and gave one half to each of the other two people. The remaining brownie was cut in four equal parts and each person received a fourth. She glued the pieces and reported her answer as $1\frac{1}{2}+\frac{1}{4}$ (see picture 1). Theodora used the same strategy and labeled the pieces with their corresponding value, 1 whole, $\frac{1}{2}$, and $\frac{1}{4}$. She reported her answer as $1\frac{3}{4}$ (see picture 2). Other students gave their answers as $1+\frac{1}{2}+\frac{1}{4}$, or as $1\ \frac{1}{2}\ \frac{1}{4}$. Slightly different strategies also led to $1+\frac{1}{2}+\frac{1}{4}$. One student solved the problem by giving one whole to each person. Then she broke the remaining 3 wholes into 6 halves and gave one to each person. She then broke the remaining two halves into fourths and gave one to each one.

Natasha also started by giving one brownie to each person. Then she divided one of the remaining brownies in four equal parts, and gave one piece of $\frac{1}{4}$ to each person. She repeated the same process with the other two brownies. She displayed her answer as $1\frac{3}{4}$ (see picture 3). Another student used a very similar strategy, but after giving one brownie to each person, she cut all three remaining brownies in four parts.

Connor decided to cut each of the seven brownies in four equal parts, so each of the parts was $\frac{1}{4}$. He realized that he had 28 such pieces, divided 28 by 4, and gave 7 pieces to each person (see picture 4). He reported his solution as $\frac{7}{4}$.

After most students had solved the problem and recorded their solution on paper, Ms. Klein asked several students to share their strategies and solutions with the rest of the class. As students described their solution verbally, Ms. Klein modeled their process on the overhead projector. The teacher represented seven brownies, and four people. As students explained what they had done with each brownie or fractional part, Ms. Klein would model that step on the overhead projector, place the corresponding piece next to each person, labeling each piece with the corresponding fraction. Once a brownie had been shared, it was crossed out from the pool (see figure 1). Other students went to the front of the classroom with their display sheet and explained their strategy and shared their answer.

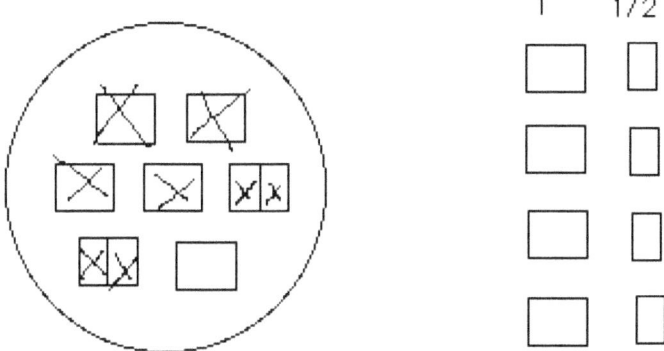

Fig. 1 First one brownie to each person; then half a brownie to each one...

Helping make connections
In the third grade or fourth grade, activities like the one described can set the stage to make explicit many important mathematical connections. One connection is the relationship between division of whole numbers and fractions. Often, students conceptualize division of whole numbers and fractions as separate and unrelated topics. A setting such as the one described, where the units can be further subdivided can help students make the connection. Consider how different the situation would be if the objects to be shared could not be subdivided; a remainder, rather than a fraction would be a more natural part of the response. For example, if 9 marbles are to be divided among 4 children, then each one would get 2 marbles, and there would be one marble remaining; it would not be possible for the children to cut one fourth of a marble.

Some children will notice that the same numbers appear in the stated division problem $7 \div 4$ and in the answer, $\frac{7}{4}$. Kristin, a fifth grader, had written the values $1, \frac{1}{2}$, and $\frac{1}{4}$ in her answer. She then converted everything to fourths and found that each person had received $\frac{7}{4}$. She looked at the numbers in the answer and in the statement of the problem. She underlined the 7 and the 4 of the original problem $7 \div 4$ and expressed her amazement, "it is just weird that it came equal to this and this" (pointing at the 7 and the 4). Some students will notice the pattern after a few examples; some will need explicit guidance to establish the connection. Teachers can help students see the relationship by guiding them to rewrite their answers in several ways, one of which will bring the connection to the forefront. A child who writes the answer of a problem like $5 \div 4$ as $1\frac{1}{4}$ may see the connection more readily if encouraged to write the answer also as $\frac{5}{4}$.

People who know the connection between division and fractions, often switch back and forth between the two concepts with ease, and sometimes without even noticing. Often the slanted fraction bar indicates division. For example, many calculators use the slanted fraction bar to indicate the division key. Some calculators that use the symbol $\div$ on the division key, use the slanted bar on the display. However, students need time and opportunity to establish that connection. The connection between $3 \div 4$ and $\frac{3}{4}$ will not be automatic for most children, nor should we assume that it is a natural connection. Some children will even resist using fractions to report a division problem. A student in another classroom stated vehemently, "We are not talking about fractions, we are talking about dividing" (Toluk 1999, p. 182).

Another connection that can be made using the brownie problem as a springboard is that of equivalent fractions. Ms. Klein displayed for the class three answers obtained using different strategies: $1+\frac{1}{2}+\frac{1}{4}$, $1\frac{3}{4}$, and $\frac{7}{4}$. The class discussed the different answers and agreed that they were all equal. Students argued that in each case, seven brownies were divided equally among four people, so the answers, even though they were written differently, represented the same amount. A student stated that "just because you have different numbers, does not mean they [the fractions] are different". Ms. Klein summarized the equivalencies on the overhead projector by writing three equations

$$1\frac{3}{4} = \frac{7}{4}, \quad 1+\frac{1}{2}+\frac{1}{4} = 1\frac{3}{4}, \quad \frac{7}{4} = 1 + \frac{1}{2} + \frac{1}{4}.$$

In this situation, students may be seeing the equal sign in a way they are not used to. Often students assume $=$ means that an operation on the left side of the sign has to be performed and the result written on the right side. Understanding the meaning of the equal sign is important, and it will be crucial for the future (in algebra for example). Teachers in earlier grades can lay a foundation for this understanding by using the equal sign to show equivalency.

Even after students have realized that the answers $1\frac{3}{4}$, $1 + \frac{1}{2} + \frac{1}{4}$, and $\frac{7}{4}$ represent the same amount, it is important to give them other opportunities to make explicit why some fractions are equivalent. For example, when stating $1\frac{3}{4}= 1 + \frac{1}{2} + \frac{1}{4}$, they need to make explicit that $\frac{1}{2} + \frac{1}{4} = \frac{3}{4}$. In third grade it is not necessary to introduce the concept of common denominator; students can realize that the fractions are equivalent by overlapping the rectangles that represent them. Students will realize that $\frac{3}{4}$ and $\frac{1}{2} + \frac{1}{4}$ do indeed cover the same amount of the unit rectangle. They can also break $\frac{1}{2}$ in two equal parts and use the fact that $\frac{1}{2} = \frac{1}{4} + \frac{1}{4}$ to make the connection. In the same way, students can realize that $1\frac{3}{4}$ and $\frac{7}{4}$ are equivalent; in each case $\frac{1}{4}$ is missing to complete two units. Or they can use the fact that $1=\frac{4}{4}$ to see that there are seven $\frac{1}{4}$ in $1\frac{3}{4}$. At this age, it is not important to develop efficient algorithms to change between mixed fractions and improper fractions. What is important is that students have ways to convince themselves and others whether or not two fractions are indeed equivalent. At this point, when comparing fractions to establish equivalency, it may also be convenient or even necessary that the teacher provides pre-cut fractions so that the comparisons are done with pieces that are exact.

Related to the concept of equivalent fractions is the idea of equal ratios. Students may realize that dividing two brownies among four people is the same as dividing one brownie between two people (see figure 2). In both cases, each person receives the same amount. Some will notice an inverse relation between the number of the pieces and their size for fractions that are equivalent. For example, when comparing a solution expressed as $\frac{2}{4}$ with one expressed as $\frac{1}{2}$, that one has twice the number of pieces as the other, but the pieces are only half as big; comparing $\frac{3}{6}$ with $\frac{1}{2}$, the number of pieces is three times but the size is only one third.

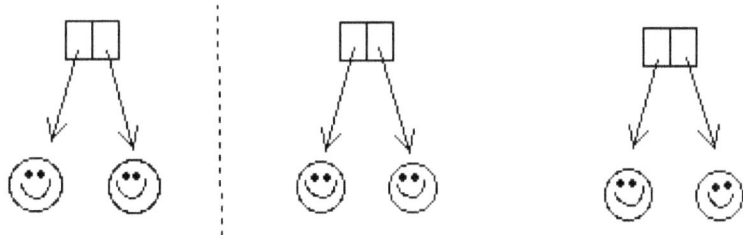

Fig. 2. $1 \div 2$ gives the same as $2 \div 4$

Students can solve problems like $2 \div 4$, $3 \div 6$, and $4 \div 8$ and realize that in each case the result is $\frac{1}{2}$, or a fraction equivalent to $\frac{1}{2}$ like $\frac{2}{4}, \frac{3}{6}$, or $\frac{4}{8}$. Some of the students will notice that the number to be divided is one half of the divisor. Some will realize that the result of two divisions is the same if in both cases the ratio between dividend and divisor is the same. So $3 \div 2$ is the same as $30 \div 20$ and as $300 \div 200$. In the upper elementary grades, this realization will help students understand the reason why to divide decimals like $2.5 \div 0.5$ we can instead solve the problem $25 \div 5$ and obtain the same answer.

In the same way, for fractions that are equivalent to $\frac{1}{2}$, some of the students will see that the numerator is one half the denominator. This realization will lay the foundation to see that fractions represent not only a quantity but also a ratio. Equivalent fractions are equal as quantities but also represent the same ratio.

Another important connection in a division problem is the relation between the corresponding fraction with the remainder and the divisor. For example, $9 \div 4$ can be reported as 2 R1; it can also be reported as $2\frac{1}{4}$. Many students need to be guided through the additional step of dividing the remainder also and obtain a fractional answer. A context where it makes sense to divide the remainder can enhance the connection. Thinking about $9 \div 4$ in terms of money, for example, students can realize that each person will receive 2 dollars and one quarter, or $2\frac{1}{4}$.

With Connor's approach described above, having seven pieces of size $\frac{1}{4}$ can be a very natural process. As Connor did, this can be written as $\frac{7}{4}$. This fraction is clearly bigger than $\frac{4}{4}$ or one unit. Thus in mathematics the word *fraction* denotes the quotient of two quantities, and it is not uncommon that the quantity being divided is bigger than the divisor. However we need to realize that in everyday language, *fraction* usually is used to denote a fragment, a part, not the whole. Often all the examples that students first encounter of fractions in school are parts of a unit. Furthermore, the unfortunate name *improper fraction* may suggest to some children that there is something "illegal" with a fraction like $\frac{7}{4}$ or that it is not proper to use such fractions. To facilitate dealing with fractions like $\frac{7}{4}$, it is important to let students view fractions such as $\frac{3}{4}$ not only as 3 out of four pieces, but also as 3 pieces of size $\frac{1}{4}$ each. For many students it is easier to extend this latter meaning to fractions bigger than the unit, while many have some problems with the "seven out of four" interpretation.

Addressing misconceptions

By the time children enter school, they have developed many informal mathematical ideas. In school they develop their own conceptions of mathematical ideas based on the examples and experiences provided by textbooks and teachers. Sometimes the conceptions that children form are based on a limited number of examples; sometimes they form conceptions unintended by the teacher. Often these conceptions will be incomplete or even be misconceptions. It is also important to remember that these preconceptions can be quite resilient and it is not enough for teacher to show the "right way."

Students will have previous conceptions about division. For example, one of the students in Ms. Klein's classroom objected to the division problem 7 / 4 because 4 does not go evenly into 7. The teacher may guide students to make sense of the measurement interpretation of division and of "how many times does 4 go into 7". Concrete representations of 7 and 4 that lend themselves to measuring one quantity in terms of the other, such as strips of paper with clearly marked segments, can be helpful. Students can see that 4 does not go two times into 7, but it goes in one time, and there is still some room left. Students can see that they can fit three out of the four segments that form 4, so that the answer is 1 and 3/4.

Even when students have the concrete representation of fractions in front of them, it is very common to operate only with the symbolic representation and forget to check whether the answer makes sense or not. A common mistake when trying to determine $\frac{1}{2} + \frac{1}{4}$ is to "add across" and obtain an answer of $\frac{2}{6}$. It is not enough for the teacher to point out that fractions are not added that way. Rather, the student could represent the problem with a piece of $\frac{1}{2}$ and a piece of $\frac{1}{4}$, and explain what it means to add the two quantities, and realize that the answer will be bigger than $\frac{1}{2}$. The student can also represent $\frac{2}{6}$ of the same unit and see that this fraction is less than $\frac{1}{2}$, so that $\frac{2}{6}$ cannot be the answer to the problem $\frac{1}{2} + \frac{1}{4}$.

Relating the problem to some more familiar problems can help students make sense of the answer. In the beginning students make sense mainly through connections to concrete objects and situations. Gradually students should develop the ability to make sense by relating a problem in mathematics to other mathematical situations they are more familiar with. For example, Bokú, a student in the upper elementary grades, when asked how he obtained the answer $1\frac{3}{4}$ for the division $7 \div 4$, said "4 times $1\frac{1}{2}$ is six, four times two is eight, so the answer has to be halfway between $1\frac{1}{2}$, and 2."

Conclusion

Given the proper context, tools, and guidance, young students can develop an understanding of fractions and make connections to other areas of mathematics. With teachers' help they can build a foundation to later deal in more systematic ways with equivalent fractions, converting fractions from one representation to another, and be able to move back and forth between division of whole numbers and rational numbers. Not all examples or contexts are equally powerful to establish connections, nor do all serve equally well to dispel misconceptions. It is clear from the variety of strategies used by students that there is no single most powerful form of presenting an idea. The teacher needs to develop a repertoire of alternative approaches and forms of presenting ideas. Teachers can derive them from several sources, including their students' strategies, research, exemplary materials, other teachers, and from reflection on their own experiences of what works and why.

References

Tierney, Cornelia, and Mary Berle-Carman. *Fair Shares*. Palo Alto, Ca.: Dale Seymour Publications, 1995.

Toluk, Zulbiye. *Children's conceptualization of the quotient subconstruct of rational numbers*. Unpublished doctoral dissertation, Arizona State University, 1999.

9 HOW DOES IT FEEL? TEACHERS COUNT ON THE ALPHABET INSTEAD OF NUMBERS[9]

Young children need time and opportunities to develop a conceptual understanding of numbers. As adults we have developed familiarity of the number sequence and flexible number sense and it is often hard to remember what it was like at the beginning when the number sequence was new for us. Just because children are proficient in reciting the numbers up to twenty or beyond, it does not guarantee that they have a sense of how big a number is, or what quantity it represents. As adults we have developed not only a sense of the size of the numbers, but also a rich network of relations among them (for example, 5 can be broken into 4+1 or 3+2). We often overlook how much knowledge and experience children need to develop the rich number concept and fluency with numbers that we have achieved over time.

In this paper, we present an activity using the alphabet that we use in pre- and in-service teacher training programs. This alphabet activity is often used in professional development programs, such as *Cognitively Guided Instruction* (see Carpenter, Fennema, & Franke, 1996; Carpenter, Fennema, Franke, Levi, & Empson, 1999 to read more about CGI) and we adapted the activity for our pre- and in-service teacher training programs. The goal of the activity is to help teachers understand what children go through as they begin to learn the number sequence and construct strategies for solving arithmetic problems. This activity is not designed for children but for teachers, so that teachers can better understand children's experience.

In the following, we discuss first counting principles that children need to acquire to develop counting knowledge with understanding and children's informal strategies for simple addition and subtraction identified in the

[9] Baek, J. M. and Flores, A. (2005). How does it feel? Teachers count on the alphabet instead of numbers. *Teaching Children Mathematics*, *12*(2), 54-59. Copyright National Council of Teachers of Mathematics. Used by permission.

literature. Then, we discuss the alphabet activity that we have used in the teacher education program.

Background
Counting Principles

Counting is more than simply reciting a string of numbers in a rote fashion. Children need to acquire conceptual meanings of number words and counting procedures. Knowledge of the following counting principles and the part-whole concept provides a basis for children to construct meaningful strategies successfully to solve arithmetic problems. The following principles have been identified as prerequisite to master counting (Baroody & Ginsburg, 1986; Fuson & Hall, 1986; Gelman & Gallistel, 1978; Gelman & Meck, 1986). We present the following principles to teachers using concrete objects and discuss what it means to learn how to count.

1. Stable order: Words used in counting must be the same string of words from one count to the next, such as "one, two, three, four,..." instead of "one, two, four, three, five, ..."
2. One-to-one correspondence: In counting objects we match each number with one object and count all of the objects.
3. Item irrelevance: The objects to be counted do not have to be the same kind.
4. Cardinality: The last number counted is the total number of objects.
5. Order irrelevance: The objects can be counted in any order and get the same result.

Research shows that mastery of the first three principles allows children to count in a correct manner and the last two principles help children construct conceptual knowledge of numeracy (Fuson & Hall, 1983; Van de Walle, 1990).

In the following section, we describe children's strategies for simple addition and subtraction, so that we can compare the strategies constructed by children and teachers.

Children's Strategies for Adding and Subtracting Single-digit Numbers

A number of studies have identified three levels of strategies that children construct to solve addition and subtraction word problems with single-digit numbers and how they evolve over time (see Fuson, 1992; Kilpatrick, Swafford, & Bradford, 2001 for summary of the research findings). Initially young children employ *Direct Modeling* strategies, using physical objects to directly represent quantities and actions or relationships in a problem. Over time, the strategies evolve to *Counting Strategies* and *Derived Facts*. We describe an example of each strategy level for adding 8 and 5 objects.

- Level 1: Direct Modeling. Children use objects or fingers to represent each of the quantities in the problem, put them together, and count all the objects. For 8+5, they make a set of 8 objects and set of 5 objects, push them together, and count them all, "One, two, three, ..., thirteen."

- Level 2: Counting Strategies. In applying a counting strategy, children do not represent both quantities in the problem, and represent only the second addend to keep track of how many are added. In solving 8+5, for example, children say, "8" and count "9, 10, 11, 12, 13" as they keep track on their fingers or objects to make sure they count five more.

- Level 3: Derived Facts. Children often learn certain number facts before others and use those number facts to derive solutions for other problems. Children usually learn doubles (e.g., 3+3, 4+4) and sums of ten (e.g. 4+6, 7+3) before other facts, and use them to solve other problems. For example, children often figure out 8+5 is 13 because "8 and 2 is 10. 3 more is 13."

Counting with the Alphabet

Even after children master the counting principles discussed above, it takes a long time and plenty of experimentation and learning opportunities to develop number sense and proficient strategies for adding and subtracting small numbers. As adults, we tend to underestimate how much understanding children need to be proficient in counting numbers and use the knowledge in problem solving situations. We often underestimate how confusing it would feel for children when we expect them to make a big leap as soon as they start to recite the number sequence. Often the fluency of children in reciting the number sequence makes us think that they have the same depth of understanding of quantities as we do.

To get a sense of what children go through as they develop number sense and problem solving strategies, we use an activity that pre- and in-service teachers have to use the alphabet sequence instead of numbers in solving arithmetic problems. Teachers are instructed not to translate the letters to numbers. We chose to use letters because we all know the alphabet sequence well, just like children know the number sequence, but it does not mean that we mastered the quantity that each letter would represent if the letters were used for counting. We present the following set of word problems on a board and provide each teacher with chips or other counting manipulatives. (The word problems are from Fennema, Carpenter, Levi, Franke, and Empson 1999. We replaced the numbers in their problems with letters for our activity.)

- Lucy has *H* fish. She wants to buy *E* more fish. How many fish would Lucy have then?

- TJ had M chocolate chip cookies. At lunch she ate E of them. How many cookies did TJ have left?
- Janelle has G trolls in her collection. How many more does she have to buy to have K trolls?
- Max had some money. He spent I dollars in a video game. Now he has G dollars left. How much money did Max have to start with?
- Willy has L crayons. Lucy has G crayons. How many more crayons does Willy have than Lucy?
- K children were playing in the sandbox. Some children went home. There were C children still playing in the sandbox. How many children went home?

Observations

It is interesting to note that the strategies that both pre- and in-service teachers use to solve problems with letters are the same *type* of strategies as young children use for problems with numbers. In the classes we used this activity, most teachers used a Direct Modeling strategy. For example, in solving the problem, "Lucy has H fish. She wants to buy E more fish. How many fish would Lucy have then?", most teachers represent each set as they count, "A, B, C, D, E, F, G, H, and A, B, C, D, E," and then count them all, "$A, B, C,..., K, L, M$ (Figure 1)." This strategy is identical to a Direct Modeling strategy that young children use.

A relatively small number of teachers used a strategy more sophisticated than Direct Modeling. For example, in solving the problem, "Janelle has G trolls in her collection. How many more does she have to buy to have K trolls?", Ms. Jacobson (all the names are pseudonyms), did not represent both sets but represented only one of the sets by drawing five counters. She said, "G", and counted, "H, I, J, K," as she was pointing each counter for each letter (Figure 2). A number of teachers represented the first set, G, and added more until they had K (Figure 3). It was an awakening discovery for teachers to see that they didn't have to represent the set G because they do not act on the first set. Ms. Jacobson's strategy is classified as a Counting strategy, which is a second level strategy that children use for addition or subtraction problems.

A few teachers tried to use "letter facts". For example, Ms. Perry knew right away that the difference between M and E would be H in the second problem because in the first problem the sum of H and E was M. However, it was interesting to observe that even though the two problems were given consecutively, very few teachers notices immediately the connection. That is, very few remembered they just had solved $H + E = M$ when trying to find $M - E$. Mr. Wilson tried to use E and J, representing 5 and 10 respectively, as benchmark letters. In solving the last problem, "K children were playing in the sandbox. Some went home. There were C children still playing in the sandbox. How many children went home?", he said, "I know that the problem is like $K - \Box = C$, and $K = J + A = E + E + A = E + C + C$. So $\Box = E + C$ which is (counting E, F, G, H) H" (Fig. 4). If we think of his strategy in numbers, Mr. Wilson was decomposing 11 into 10 and 1, and again into 5, 5, and 1, and again into 5, 3, and 3. Many teachers in the class did not understand what he was trying to do first. After some discussion, they understood, and realized that developing derived fact strategies by breaking down a number into smaller and more manageable size of numbers is not a simple task when you are developing concepts of numbers. It is interesting to note that Mr. Wilson picked 5 and 10 as benchmark numbers, just like a lot of young children do in primary grades. Teachers also discussed about how frustrating it would be if they were instructed to memorize facts, such as $H + E = M, P - I = G$, or $K - H = C$ when they do not have a solid understanding of numbers.

Teachers' reactions

In one of the courses, prospective teachers were asked to select five items from the work they produced during the semester and create a portfolio. The goal was to include items that helped *growth* in their understanding of how to teach mathematics. One of the preservice teachers included the activity with the alphabet in her portfolio. Her explanation of why she included it was

> to exemplify how students may feel when learning a new concept. This activity was very different than any I had encountered before. It was a great activity to show me how students may think, act, and feel when learning a new concept.

Several teachers commented that the problems were indeed much more difficult to solve using the alphabet than using the corresponding numbers. They began to appreciate the power of Direct Modeling strategies that young children use to solve simple addition and subtraction. The teachers also noticed that they do not automatically memorize "letter facts" like $E + C = H$ just by working on several problems. It helped them to be more aware of the fact that children need many opportunities to internalize the corresponding number facts. It was also interesting to see that after going through the process themselves, teachers paid more careful attention to the strategies that children use.

Conclusion

This is not an activity that we suggest teachers to use with their students. This is an activity that could help teachers understand what children go through when they are at the beginning stage of developing number concepts and strategies for addition and subtraction. Because teachers have developed solid understanding of whole numbers over a long period time, it is very easy to underestimate the difficulty that young children experience at the beginning. It aims to help teachers understand the importance of time and opportunities that children need to develop fundamental concepts of numbers and adding/ subtracting strategies. The reactions from participant pre- and in-service teachers after the instruction are a confirmation that this alphabet activity could help them to develop a better understanding of children's learning process of numbers and arithmetic operations and to rethink instructional implications for young children. Teachers also expressed more sensitivity to recognizing the developmental level of the child. Often teachers want children to use more efficient strategies, such as counting on, rather than modeling and counting from the beginning. Teachers have often expressed they perplexity how some children, even after having been taught to count on, revert to direct modeling, as soon as the instructor leaves. The alphabet activity was a reminder for teachers of the complexity of the process to learn counting strategies and derived facts strategies.

References

Baroody, Arthur J., and Herbert P. Ginsburg. "The Relationship between Initial Meaningful and Mechanical Knowledge of Arithmetic." In *Conceptual and Procedural Knowledge: The Case of Mathematics*, edited by James Hiebert. Hillsdale, N.J.: Lawrence Erlbaum Associates, 1986.

Carpenter, Thomas P., Elizabeth Fennema, Meagan L. Franke, Linda Levi, and Susan B. Empson. *Children's Mathematics: Cognitively Guided Instruction.* Portsmouth, NH.: Heinemann, 1999.

Fennema, Elizabeth, Thomas P. Carpenter, Linda Levi, Meagan L. Franke, and Susan B. Empson. A Guide for Workshop Leaders: Children's Mathematics Cognitively Guided Instruction. Portsmouth, NH.: Heinemann, 1999.

Fuson, Karen C., and J. W. Hall. "Acquisition of Early Number Word Meanings: A Conceptual Analysis and Review." In *The Development of Mathematical Thinking*, edited by Herbert Ginsburg. New York: Academic Press, 1983.

Fuson, Karen. "Research on Whole Number Addition and Subtraction." In *Handbook of Research on Mathematics Teaching and Learning*, edited by Douglas A. Grouws, pp. 243-275. New York: Macmillan, 1992.

Gelman, Rachel, and C. R. Gallistel. *The Child's Understanding of Number.* Cambridge: Harvard University Press. 1978.

Gelman, Rachel, and Elizabeth Meck. "The Notion of Principle: The Case of Counting." In *Conceptual and Procedural Knowledge: The Case of Mathematics*, edited by James Hiebert. Hillsdale, N.J.: Lawrence Erlbaum Associates, 1986.

Kilpatrick, Jeremy, Jane Swafford, and Bradford Findell. *Adding it Up: Helping Children Learn Mathematics*. Washington DC: National Academy Press, 2001.

Van de Walle, John A. "Concepts of Number" In *Mathematics for the Young Child*, edited by Joseph N. Payne, pp. 63-88. Reston, VA: National Council of Teachers of Mathematics.

10 WRITING FOR AN AUDIENCE IN A MATHEMATICS METHODS COURSE[10]

The first time many prospective elementary teachers make explicit and confront previous conceptions about mathematics and its teaching is during their mathematics methods course. This makes this course a crucial experience in their evolution to become teachers of mathematics. Prospective teachers in a mathematics methods course need to develop the ability to reflect on their actions, beliefs, knowledge, and attitudes. Writing in a mathematics methods course fosters reflection in a natural way; it provides future teachers with a tool for documentation, analysis, and discussion to help them reach new comprehension. At the same time, what teachers write provides us with a window into their reflection and growth process. In a previous paper, we described the benefits that writing in mathematics has for students, as well as the benefits that writing in a mathematics methods course has for prospective elementary teachers (Flores & Brittain 2003). We also described some of the writing tasks used with prospective teachers. In this paper we will focus on the process of writing and the importance of writing for an audience in a mathematics methods course for elementary teachers. The writing examples presented here were collected in several one-semester courses taught by the first author. The second author participated as a student in one of the courses.

The process of writing
We will first discuss different kinds of writing. Britton and his colleagues

---

[10] Flores, A. and Brittain, C. M. (2004). Writing for an audience in a mathematics methods course. *Teaching Children Mathematics*, *10*(9), 480-486. Copyright National Council of Teachers of Mathematics. Used by permission.

categorize mature writing according to function (Britton, Burgess, Martin, McLeod, & Rosen 1975). The categories are transactional, expressive, and poetic. The writing that was done in the mathematics methods course falls in the two categories expressive and transactional. We will give Britton et al.'s description of transactional and expressive writing. *Transactional writing* is language to get things done: to inform people (telling them what they need or want to know or what we think they ought to know), to advise or persuade or instruct people. Thus the transactional is used for example to record facts, exchange opinions, explain and explore ideas, construct theories; to transact business, conduct campaigns, change public opinion. (Britton et al., 1975, p. 88).

Within the informative function of transactional writing we find recording, reporting, as well as writings with increasing levels of generalization.

Expressive writing includes, among others, the kind of writing that might be called 'thinking aloud on paper', which is intended for the writer's own use; the kind of diary entry that attempts to record and explore the writer's feelings, mood, opinions, preoccupations of the moment (Britton et al. 1975). Writing addressed to a limited audience that shares much of the writer's context and many of his or her values and opinions and interests also has an expressive function. Britton et al. make three generalizations about the expressive function.

Firstly, expressive language is language close to the self. It has the functions of revealing the speaker, verbalizing his consciousness, and displaying his close relation with a listener or reader. Secondly, much expressive language is not made explicit, because the speaker/writer relies upon his listener/reader to interpret what is said in the light of a common understanding (that is, a shared general context of the past), and to interpret their immediate situation (what is happening around them) in a way similar to his own. [...] Thirdly, since expressive language submits itself to the free flow of ideas and feelings, it is relatively unstructured. (Britton et al., 1975, p. 90)

In developmental terms the expressive plays a crucial role, according to Britton et al. (1975). From expressive writing more differentiated forms of mature writing are developed as students grow. Also, for mature writers many times a first draft of a transactional writing is essentially expressive.

The process of writing to learn in mathematics

Because writing is a tool to learn, it is more important to ask whether using writing will improve students' learning of mathematics, rather than asking whether writing in a mathematics class will improve students' writing abilities (Connolly & Vilardi, 1989). It is possible that increased opportunities to write might actually increase the writing skills of the students, but for the mathematics teacher, the goal is to use writing to help students understand mathematics (Countryman, 1992). Students write because of what they learn in the process itself, not because of the final product (Gribbing, 1991). Writing is just the "aide" to make content more meaningful and clear for the students, not the goal in itself. In the same way, prospective teachers' writings should be looked not as final products where correctness, elegance, cohesiveness, structure are important, but as tools that promote teacher change.

We recommend the use of process writing when using *writing to learn*. Process writing is a successful and well-documented approach to foster writing (Peregoy & Boyle, 1993; Freeman & Freeman, 1995). In process writing, students approach writing as a series of five interrelated phases – pre-writing, drafting, revising, editing, and publishing (Peregoy & Boyle, 1993).

In a simplified version of process writing, Peterson and Eeds (1994) propose a three-phase sequence – fluency, content, and correctness. This model could be appropriate for mathematics teachers in the integration of writing in their classroom. Students first focus on fluency. They are encouraged to write as they think without concentrating on mechanics or spelling. Students write as much as they need to describe their thoughts or express their opinion about the subject. Second, once a first draft is done, students must engage in a dialogue with their peers and teachers by sharing their writing and asking for feedback about the content. This dialogue between the student and her peers and teacher allows her to negotiate meaning and make sure that whatever was written makes sense to both the author and the audience (e.g. Do you receive the message that I intended? Is it clear what I mean?). Finally, once meaning has been negotiated and the writer and the audience agree on what the text says, they can focus on cleaning up the piece and check for misspellings and mechanics.

Writing in the mathematics methods course

Prospective teachers write frequently and consistently during the semester. As one of them put it: "I did more writing in this math course than I have done in some of my English classes." Prospective teachers *write to learn*. The main purpose of writing in this methods course is to help teachers grow in their knowledge, and to change their beliefs and attitudes when necessary. As with younger students in content-area classes, frequent short writing assignment that are collected and responded to on a regular basis serve this purpose better than one or two 10- or 20-page long term paper (Kneeshaw 1992).

Because the writing is considered as a means and not as an end in itself, correctness of the writing (spelling, syntax, structure, etc.) is not important as long as the reader is able to understand without undue effort what the writing is communicating. The emphasis is on the flow of ideas and feelings into the text. Samples of future teachers' writing included here are therefore not finished and polished examples of their writing skills. Rather, those writings are ways they have to let their ideas and feelings flow and help them learn how to become teachers, and at the same time provide us with a picture of how they change.

Most of the writing that serves as a tool to learn in this course would fall into the category of expressive. In the case of the few writings (such as the midterm, reports) whose main purpose is to show what prospective teachers have learned (which may be classified therefore as transactional writing) also serve as a learning tool because they help them organize and clarify their ideas.

Writing for an audience

Other than taking class notes, students usually do not write spontaneously in a mathematics methods course. Writing tasks are explicitly assigned. In the beginning, prospective teachers are not comfortable writing in a mathematics methods course. For the most part they had never expressed their feelings, opinions, ideas about mathematics in writing.

"I never related writing to math."

"I always thought of writing and math as being two parallel skills that never met."

"In all the other mathematics courses that I have taken, I have never had to write anything but the equation that I am planning to answer."

"Writing is perceived as an action that doesn't usually take place in a math course."

"Incorporating written assignments into a mathematics class is a very new concept to me. I have always experienced a numerical based mathematics class rather than written assignments and activities."

The first perceived audience of the writings in the course is usually the professor. In order to make writing a tool for learning and discovery, it is important that the professor is perceived as someone to be communicated with, rather than someone who reads their writings for grading purposes. It is important that every writing collected is responded to in very specific terms. Usually a couple of sentences are enough, although occasionally some of the students' questions require lengthy responses. As one student said, "writing in this class taught me that I was valued as a student. Your comments showed that you actually read and were interested in what I wrote. ... Writing established a dialogue." Another student expressed, "an important part of the math writing was that there was always going to be a response."

One important element of the use of writing in the methods course is that the professor is perceived as someone interested in communication, rather than assessment (Johnson et al, 1993). By responding to all the writing assignments, the professor helps future teachers to understand this role. One student wrote in her final portfolio about one of the early reactions:

I was still very unaccustomed to writing about math …. I had just begun to realize that our professor indeed read everything we wrote and always replied. There was a dialogue. Again, something I had not experienced.

When the professor writes comments back to his/her students, he/she participates in the dialogue. Then, the instructor becomes part of the audience:

I find these reflections beneficial because it gives me a way to let you know what point I am at and it also is my way of writing down what I thought helped me and what I thought didn't quite work. I think the reflections are great because it is a way that I know that I can show you that I am either confused or progressing in math.

The fact that the professor always responds makes future teachers perceive their writing as part of a dialogue, and they perceive the professor as someone they can trust their thoughts and feelings about mathematics learning and teaching. As Britton et al. point out, "a writer who envisages his reader as someone with whom he is on *intimate* terms must surely have very favourable conditions for using the process of writing as a means of exploration and discovery" (Britton et al. 1975, p. 82).

This process does not happen overnight. Some of the writings in the beginning reveal little of the student's thinking, beliefs, and feelings. With time, and with some explicit encouragement, the writing changes from being just summaries into texts that serve as windows into prospective teacher's thinking and feelings. One student describes this change:

I have never had to express myself in writing when dealing with mathematics. When we first began to write the reactions I wrote very summary-like, not going into a lot of my opinions. I know now that it was because I thought it was just for a grade and I didn't see the importance behind it. As we wrote more and more reactions, I began to take my time on them and really express what I had learned. I finally realized that I was "allowed" to write about what was important to me and there wasn't a right or wrong answer to my math assignment like there had always been before.

This is from one student who wrote a summary: "When you returned my paper I was surprised to read that you wanted to know what I thought about what I had read and not only what the authors said."

Writing provides a non-threatening way to communicate. As one of the students expressed, "the medium of pen and paper gave me the sense of safety and distance that I needed in order to be heard." In addition, having the professor as a trusted audience, and knowing that the communication is individual, helps students to get the security to express in their writings feelings and ideas that they would be hesitant to express in the classroom. In this course this individual dialogue played a fundamental role. This is what some of the students wrote with respect to this issue:

"Writing gave me a voice, a way to ask a question I might not have felt comfortable in class to ask."

"When we have just done something exciting in class, I get to express my feelings. Without the fear of anyone's comments, I can express myself in regard to anything that excites me in mathematics."

"In writing, I can express my feelings that I wouldn't say in front of other students / peers. I would ... rather (at times) discuss my feelings of math individually."

"Writing also allows the student to privately communicate with the teacher about issues they might be too embarrassed to say in front of the class."

"I think that if I knew that someone other than you was to read my writings, that I would write in a completely different manner. ... I think that it takes away from the personal reflection."

Some of the writing done in the course is meant to be shared with peers. In that case it is always clear before students write something that it is going to be shared. Some teachers found beneficial to share some of their writing with their peers. The importance of sharing is that it allows student to negotiate meaning and expand their knowledge. Part of the drafting stage of the writing process involves "sharing" for negotiation of meaning, making sure that the audience understands the message as intended (Peterson & Eeds, 1994). Sharing might also provide an opportunity to learn from peers who may have a different perspective about a common concept or learning situation. In this course, teachers had the opportunity to work in pairs and small groups. Each member of the group wrote a short strategy about how to solve a teaching problem. Some students found that writing forced them to think on how to communicate their ideas to others. Then the members shared their approaches to engage in a group discussion. When students interacted to do the activity and to negotiate what would be reported, sharing of ideas and clarification took place. This sharing allowed the teachers to see their peers' perspectives. As one pre-service teacher explained: "Today, I realized that the way that seems so obvious to me is not necessarily the obvious one to others." After the discussion, a general report was produced. Sometimes, all the approaches were included and sometimes the teams came up to a consensus and just one approach was selected.

Of course, not all writing in mathematics is narrative. For example, students worked in small groups and were asked to prepare a poster that summarized what were the salient aspects when teaching the meaning of operations to students. This poster became an important piece of information, not only for the authors, but also for the other pre-service teachers in the classroom. This poster is an example of writing for authentic purposes and publication, two concepts that have been promoted by educational research as important factors that foster writing in students (Peregoy & Boyle 1993).

As the semester progresses, prospective teachers realize what every person who keeps a personal journal knows, that the writer is also the audience. "The class reaction papers we wrote after class periods taught me how to take a few moments after learning a new method of doing things to listen to what your mind says and let it construct what knowledge is important to understand." Some of them start to use writing in a conscious and explicit way to help themselves organize their ideas, help their own understanding, clarify and elaborate ideas about mathematics and its teaching, release tension, express their emotions, etc.

Another benefit that some teachers expressed was that writing allowed them to learn about themselves and their growth and changes over the semester. The written record let them contrast their positions at different points in time. At the beginning of the semester, one of the students wrote in her biography: "math is one of the few subjects which is absolute -- the answer is right or wrong and the feedback is immediate." At the end of the semester, she reflected how her perception had changed:

My training in mathematics as an elementary and high school student geared me toward this kind of thinking. Since this class, I think that mathematics includes different ways of thinking about math problems and ways to solve problems. I can see that not all people look at math problems the same and that different perspectives come into play.

Another teacher reflected on how the meaning of *understanding mathematics* had changed for her, from the first session to the end of the semester:

I said that understanding math comes fairly easy to me; however, after exploring concepts relationally, I know that I never really understood math at all. I simply could memorize formulas and plug numbers into them with no idea of why the formula worked. I had only instrumental understanding of math.

Skemp (1987) recognized "the great psychological difficulty for teachers of reconstructing their existing and longstanding schemas" (p. 161). Writing was a way to document the difficulties encountered by these pre-service teachers in the road to change their way of thinking about mathematics. Part of the difficulty is that one semester barely seems enough:

I feel like even though I have a solid foundation of mathematical concepts, I am continually reconstructing the way that these are to be taught. The difficulty is that the way I am being trained to teach is a way that I have been taught only for one semester.

One pre-service teachers could look back at her writing and reflect on her resistance to change:

In the beginning, it was difficult for me to go back to concrete concepts and explain them.... [It] was hard for me to look away from what I was taught in mathematics and open my eyes to a better way.

One of the activities that best captured the attitudes of pre-service teachers towards writing was the building of a portfolio. At the end of the semester, students selected five items from the work they produced during the course to create a portfolio. The goal was to include items that showed *growth* in their understanding of how to teach mathematics. The process of reading and collecting their writings allowed them to look back at the semester as a whole and realize how they changed:

As I went through all of my work to select the five items to be included in my portfolio … I found myself really enjoying the reflection on where I started and where I've come. It was interesting to review my work, feelings, thoughts, and beliefs as they developed, grew, and changed over the semester. I was surprised by a few things, mainly how my belief and attitude toward teaching math was changed. I'm going to enjoy the challenge now and I feel much better knowing that it's more important to find out what's meaningful to and for the student than to push skill, drill and rote paper and formula exercises.

Conclusion

By providing children with opportunities to use speaking, reading, and writing for authentic purposes, they learn to value the functional uses of language skills in their lives. Likewise, pre-service teachers became involved in the writing tasks because they found a functional or authentic purpose for the writing: All the writing activities had the potential to be implemented in a real classroom. But most importantly, the writing assignments in the mathematics methods course had the purpose to establish a *dialogue* between the pre-service teachers and the professor to expose ideas, concerns, and suggestions about how the course was fulfilling the needs of each pre-service teacher. The written dialogue between professor and future teachers served the authentic purpose of establishing a genuine communication to help the professor adjust the course to their needs. Writing in the mathematics methods course allowed these pre-service teachers to share their own experiences, positions, and attitudes in mathematics learning and receive feedback.

REFERENCES

Britton, James, Tony Burgess, Nancy Martin, Alex McLeod and Harold Rosen. *The development of writing abilities (11-18)*. London: Macmillan, 1975.

Connolly, Paul and Teresa Vilardi (Eds.). *Writing to learn mathematics and science*. New York: Teachers College Press, 1989.

Countryman, Joan. *Writing to learn mathematics*. Portsmouth, NH: Heinemann, 1992.

Flores, Alfinio and Carmina M. Brittain. "Writing to Reflect in a Mathematics Methods Course." *Teaching Children Mathematics, 10*(2) (2003)

Freeman, Yvonne S. and Freeman, David. E. *Whole language for second language learners*. Portsmouth, NH: Heinemann, 1992.

Gribbing, William G. Writing across the curriculum: Assignments and evaluation. *The Clearing House*, *64* (1991): 365-368.

Johnson, Julie M., Melinda Holcombe, Gloria Simms, and David Wilson. "Writing to learn in a content area." *The Clearing House*, *66*(3) (1993): 155-156.

Kneeshaw, Stephen. (1992). "KISSing in the history classroom: Simple writing activities that work." *The Social Studies*, *83*(4) (1992): 176-179.

Peregoy, Susanne F. and Boyle, Owen F. *Reading, writing, and learning in ESL: A resource book for the K-8 teachers*. White Plains, NY: Longman, 1993.

Peterson, Ralph and Eeds, Maryann. *Grand conversations: Literature groups in action*. New York: Scholastic Books, 1994.

Skemp, Richard R. *The psychology of learning mathematics*. Hillsdale, NJ: Lawrence Erlbaum Associates, 1987.

11 WRITING TO REFLECT IN A MATHEMATICS METHODS COURSE[11]

ABSTRACT. We will present first a brief review of the applications of writing to learn mathematics in schools, as a tool for fostering students' learning and comprehension, and discuss briefly the process of writing to learn. In the second part, we describe how writing can be used to help pre-service elementary teachers reflect to improve their understanding of the issues involved in the teaching of mathematics and to help them reflect on their own growth as teachers of mathematics. Four main benefits and uses of prospective teachers' writing were identified: contextualizing experiences in learning mathematics, an organizing tool, a reflection tool, and taking the pressure off. Writing for an audience was also an important aspect. To illustrate the points discussed we will present some examples of the writing of pre-service teachers in a mathematics methods course.

INTRODUCTION

During the last decade, several authors have highlighted the benefits for students of writing to learn mathematics. This paper will present first a brief review of writing to learn mathematics. Although there are multiple benefits for students of writing to learn mathematics, it is unlikely that teachers will use this tool unless they have had the experience themselves of writing in relation to mathematics. Through the second and main part of this paper we invite the readers to consider another a possible use of writing—a tool to

---

[11] Flores, A. and Brittain, C. M. (2003). Writing to reflect in a mathematics methods course. *Teaching Children Mathematics*, *10*, 112-118. Copyright National Council of Teachers of Mathematics. Used by permission.

help pre-service teachers reflect on their own growth as they learn to teach of mathematics. We propose that writing could be used as a reflecting tool for pre-service elementary teachers to improve their understanding of the issues involved in the teaching of mathematics. We will present some examples of the writing of pre-service teachers in an undergraduate mathematics methods course. The writing samples presented here were collected during several one-semester courses.

WRITING IN THE MATHEMATICS CLASSROOM

A tool for fostering students' learning and comprehension
Writing to learn implies the inclusion of writing assignments to help students understand, retain, analyze, and organize mathematical concepts. At the various levels of comprehension and understanding of mathematical concepts,— e.g. introduction, acquisition, application, generalization— Countryman (1992) suggests that writing can be used to help students learn the concept more effectively, develop critical thinking, and problem solving skills. By using writing, students have a permanent record of their thoughts and can go back to reflect on them. One example is the *learning logs*, where students record in their own words what they learned at the end of the day. These learning logs have been effective in helping students to understand, retain and become critical of the content (Countryman, 1992; Kneeshaw, 1992). One of the factors in the success of the learning log is that writing allows students to organize ideas, develop new applications for knowledge, or solve problems.

Writing can help students become active participants in their own learning. Writing allows students to engage in an interaction with the subject or the content area. The writing activities become a "silent dialogue" between the student and the content area where knowledge is internalized and then articulated in the learning process. Mett (1989) suggests that writing allows students to establish a personal connection to the new concepts. When students write about a subject, they are involved in an active intellectual process where they decide what is important, what is meaningful or relevant to them (Kneeshaw, 1992; Johnson et al, 1993).

It is important, however, that the teacher clarifies that writing is for the student to use as a tool to help him understand the concepts better. Treating writing as a tool and not as an assignment to be graded helps students to feel more comfortable with using writing in the mathematics classroom (McIntosh, 1991).

Writing also serves an affective purpose (Johnson et al, 1993). Classroom research has shown that when students use writing as a tool for thinking, they also gain self-understanding and confidence in dealing with their concerns. Writing may also help to personalize the subject matter by giving the students choices to apply their knowledge into areas that interest them.

Writing has also been used as an assessment tool. Since writing provides a permanent record of students' thoughts, it might be a way to document students' learning (McIntosh, 1991). Writing as an assessment tool can be used by both the teacher and the student. Students can see and reflect on their own learning by going back to their writing. Once the writing is complete, students have a permanent record of their learning that can help them to revise information or expand their application of knowledge in the future. Drake and Amspaugh (1994) found that writing can help teachers in a number of ways such as:

- diagnosing error patters;
- giving teachers insights about where instruction should begin;
- providing evidence of where and why a student has failed to make connections; and
- giving insights regarding beliefs and attitudes students hold about mathematics.

The process of writing to learn

Proponents of writing to learn suggest that short and simple writing activities are appropriate in content-area classroom, such as math, science, or social studies. Ten or twelve short exercises around the concepts of the class are more effective than one or two 10-20 page papers (Kneeshaw, 1992). Because writing is a tool to learn, it is more important to ask whether using writing will improve students' learning of mathematics, rather than asking whether writing in a mathematics class will improve students' writing abilities (Connolly & Vilardi, 1989). It is possible that increased opportunities to write might actually increase the writing skills of the students, but for the mathematics teacher, the goal is to use writing to help students understand mathematics (Countryman, 1992). Students write because of what they learn in the process itself, not because of the final product (Gribbing, 1991).

We recommend the use of process writing when using *writing to learn*. Process writing is a successful and well-documented approach to foster writing (Peregoy & Boyle, 1993; Freeman & Freeman, 1995). In process writing, students approach writing as a series of five interrelated phases – prewriting, drafting, revising, editing, and publishing (Peregoy & Boyle, 1993). In a simplified version of process writing, Peterson (1994) proposes a three-phase sequence – fluency, content, and correctness. Peterson's model could be appropriate for mathematics teachers in the integration of writing in their classroom. Students first focus on fluency. They are encouraged to write as they think without concentrating on mechanics or spelling. Students write as much as they need to describe their thoughts or express their opinion about the subject. Second, once a first draft is done, students must engage in a dialogue with their peers and teachers by sharing their writing and asking for feedback about the content. This dialogue between the student and her peers and teacher allows her to negotiate meaning and make sure that whatever was written makes sense to both the author and the audience (e.g. Do you receive the message that I intended? Is it clear what I mean?). Finally, once meaning has been negotiated and the writer and the audience agree on what the text says, they can focus on cleaning up the piece and check for misspellings and mechanics.

PRE-SERVICE TEACHERS' WRITING IN A MATHEMATICS METHODS COURSE

Modeling writing experiences

One of the goals of a mathematics methods course is to model and give pre-service teachers the same kind of experiences that they could present to their future students. The purpose of integrating writing activities in the course was to empower pre-service teachers in their learning by providing a tool to make sense of the new knowledge presented. We think that the writing that was produced during this mathematics methods course helped pre-service teachers reflect on how children learn mathematics and how writing can be an effective tool to teach mathematics in a meaningful way. (The rest of the article is about pre-service teachers; to avoid repetition, sometimes they will be referred to simply as teachers.)

The writing activities were integrated to accomplish the goals of the course. One important decision was to determine what kind of writing could help pre-service teachers become effective teachers of mathematics. What kind of writing fosters growth and positive changes in teachers' knowledge, beliefs, and attitudes towards mathematics and mathematics teaching? What writing fosters reflection and inner dialogue? To answer these questions, we focus on the role of a methods class.

Pre-service teachers enroll in a methods course with the vision that one day, they will be the teachers. They view activities in the course as opportunities to think about their future classrooms. They collect worksheets, create portfolios of activities, materials, and other artifacts that they consider they might use in their own classroom. We think that writing could be a powerful tool to internalize and personalize the lessons and enhance learning opportunities.

Every week, a new concept was introduced in the methods course (e.g. how to teach decimals, or how to use manipulatives to divide fractions). Pre-service teachers were asked to write in their journals at the end of the session about their impressions on the lessons and/or methods used. In the future, these journal entries may become a valuable collection of their past expectations and important material to reflect on.

Teachers in training can become used to writing as a tool for reflecting. By involving pre-service teachers in writing activities around math lessons, they might experience what their future students might experience in writing. For example, during the course, these pre-service teachers participated in a number of writing exercises that were tied to a mathematical activity or lesson. Pre-service teachers, in a sense, acted as elementary school students. By asking the pre-service teachers to do a writing activity appropriate for an elementary school student, the pre-service teachers experienced the process and applications of writing in a similar way as their students might encounter. Of course, not all writing in mathematics is narrative. For example, students worked in small groups and were asked to prepare a poster that summarized what were the salient aspects when teaching the meaning of operations to students. This poster became an important piece of information, not only for the authors, but also for the other pre-service teachers in the classroom. This poster is an example of writing for authentic purposes and publication, two concepts that have been promoted by educational research as important factors that foster writing in students (Peregoy & Boyle, 1993).

Given the diversity of writing abilities, even among college students, we suggest to be flexible on the uses of writing. Some students might find narratives to be more effective than short paragraphs, while others might prefer to write captions to diagrams and pictures rather than using only linear verbal narrative.

The following table briefly describes some of the options used in the methods course. Ideas for writing assignments were adapted from recommendations from several authors.

TABLE 1

Writing assignments for prospective teachers

Activity	Implementation	Benefits
Mathematical auto-biography	First writing in course. Teachers write freely about their own mathematics learning experience.	Helps to set the tone for the writings of the semester; to make writing personal and relevant.

Reflection on class activity	Teachers write, separate from class notes, reactions about the lesson of the day, readings; ask questions.	Teachers internalize learning, organize, prioritize, personalize content (how does this help me)
Reaction to chapter reading	Teachers express their thoughts, feelings, points of views about the chapter (not a summary).	Teachers focus on reading, relate to personal experience, contrast textbooks ideas with own.
Collaborative report	Teachers work in teams to write about their ideas about teaching a topic.	Teachers negotiate meanings, appreciate different points of view, understand other ways of thinking.
Pre test exercise	Teachers write for fifteen minutes. They use peer review, and then write for another fifteen minutes	Teachers organize ideas, rehearse writing ideas, engage in a dialogue that helps them understand.
Midterm, final exams	Teachers explain concepts or problems in the teaching and learning of mathematics.	Teachers learn to synthesize and illustrate what they learned.
Story problems	Teachers write story problems to be solved using the concepts currently under study or previously studied	Helps teachers understand and illustrate different meanings of operations.

Teaching presentation	Teachers write a lesson plan, carry out the lesson in a classroom, evaluate the lesson, receive feedback from placement teacher.	Teachers organize and sequence teaching ideas, plan use of materials, evaluate learning and teaching.
Observing Teaching Standards	Teachers write examples of how different standards are implemented in the course.	Makes teachers aware of underlying standards for course activities and strategies.
Children's truths in mathematics	Teachers ask children facts they consider to be true, and explain how they know they are true.	Makes teachers aware of children's thinking in mathematics. Allows future reference and analysis.
Portfolio explanations	Teachers choose five items of their work during the semester and explain why they chose those particular items.	Teachers look back at the semester as a whole. They see how they have grown and reflect on the changes they have gone through.

Some benefits from writing to reflect

From what we collected via pre-service teachers' reflections, journal entries, and portfolios, we have identified four major benefits of writing to reflect within the context of learning how to teach mathematics. These are 1) providing a context to make the learning of teaching mathematics relevant to the future teacher; 2) writing serves as a tool to organize the thoughts of prospective teachers about issues related to the teaching of mathematics; 3) writing allows teachers to be able to look back at their own thoughts and reflect on their own growth as teachers; and 4) as teachers address affective issues in their writing, it helps to take the pressure off when dealing with difficult issues of mathematics and its teaching. We will discuss these benefits in the following sections.

Contextualizing math concepts: Making mathematics more meaningful

One of the benefits that pre-service teachers expressed was that writing in the mathematics methods course allowed them to create a frame of reference to make the experience of learning to teach mathematics meaningful to them. We believe that when teachers write about new concepts or ideas, they actually engage in an internal dialogue. In this dialogue, teachers try to express in their own words how this new experience or concept applies to their own reality. This way, prospective teachers take responsibility for their own learning by deciding what is important to them, and how they can expand the new learning to their own specific situation.

The element of *choice* has been a factor that many authors consider important in fostering students learning in many other areas such as reading and language arts (Peregoy & Boyle, 1993; Peterson, 1994; Freeman & Freeman, 1995). The element of choice can also apply to the learning of mathematics. For example, as the very first activity in the mathematics course, pre-service teachers wrote a mathematical autobiography. After that, samples of biographies from past semesters were shared. This allowed the teachers to be exposed to a wide range of attitudes and experiences, many of them similar to their own. This initial assignment set the tone for the writing for the rest of the semester. In other assignments such as chapter reactions and in-class reflections, pre-service teachers frequently used biographical references. The use of biographical references allows students to contextualize and personalize the material to their own experiences. For example, when reacting to a chapter in the textbook, a teacher compared the recommendations with her own experience: "When I went to school, math was not fun or exciting. Hands-on experiences, working in groups, and discussing possibilities was cheating." Reacting to a chapter on problem solving, another future teacher wrote "I can only wish that my teachers had looked at problem solving as a process and an opportunity to discover." During the course, teachers frequently found themselves contrasting the readings and class experiences with their own long-held beliefs about mathematics: "I always thought that there was only one right answer and only one right way to do it."

We realized when analyzing pre-service teachers' writings that many of them were critical readers as they reflected on their own experiences. They used their own experiences as a frame of reference to become critical of the material or the book chapters:

"As a child I hated number lines because they we were always told to use them so they could help us. My number lines were more of a hindrance because they seemed too confusing with too many arrow being drawn from one place to another and my spacing was always off, making the number line never seem just right. Therefore, I had to rely on other means such as my fingers."

In some cases, students' experiences actually validated the readings:

"It was not until I was in high school that I had a teacher who believed in cooperative learning. Mr. Craig always included us (students) in the math problems and curriculum. He let us come up with different ways of doing problems. We were encouraged to voice our opinions on different concepts.... For once in my life, I had finally really enjoyed learning about math."

This student's reflection validated the book's assertion of the use of cooperative learning as a good approach to learning math. Writing allowed these students to use their own personal experience to understand the teaching and learning of mathematics, making the content of the course more relevant and closer to their own reality as users, learners and teachers of mathematics.

An organizing tool

Another benefit that the pre-service teachers' writing revealed was that writing helped them to organize their thoughts and questions throughout the learning process, from introduction to the concept, to assessment and evaluation.

"Before the course, I was able to solve a mathematical problem by knowing the right formula or procedure. Usually my teachers gave these formulas to me in a list. After using writing in my math class, I was able to understand the meaning behind those formulas. I was able to create a meaning from the lesson or activity by blending the new information with my prior experience."
—CMB

The midterm and final exams consisted of essay type answers to problems in the learning and teaching of mathematics. Pre-service teachers had to describe how they would address a teaching situation in mathematics. One teacher commented on the mid-term:

"We had to explain... how we would go about doing things as teachers, for instance, teaching the meanings of multiplication, as well as demonstrate our knowledge for the material. I found the midterm to be extremely helpful because it gave me a chance to organize my ideas and asses my own knowledge."

The mid-term was perhaps the most challenging activity of the course. Teachers learned during this experience how they needed to be explicit and not take for granted that their own thinking and understanding would be the same as their future students' would be. A prospective teacher expressed how difficult is the transition from a personal understanding of the concepts and skills to be taught to a comprehension of how to help students understand these ideas:

"My inconsistency in my mathematical understanding and the expression of it in essay form is apparent. I can solve problems through trial and error and I remember equations pretty well. My weakness is wrapped in the methodology and my inability to express the process that I perform."

This future teacher's reflection depicts that she realized that she was unable to articulate in a clear form the mathematical thinking she was going through when performing a mathematical operation. This is an important point to be aware for any teacher because many times teachers understand within their own frame of reference and thinking a process, but they are unable to articulate it in a clear form for the students.

We are not proposing that only teachers who are able to articulate their thoughts in writing will be able to make the material comprehensible for their students. The pre-service teachers who might have difficulty with the writing assignments might have other important skills that could become an asset to the classroom. Writing is just a tool for learning, and like any other learning tool it might be very effective for some students, while others might have difficulty with it. Further, like any other skill or learning tool, it takes time and practice for students to feel comfortable with writing to a point that they can actually take advantage of writing to reflect. Writing about mathematics may not be easy for many prospective teachers because they are not used to it. One student expressed: "It was hard for me, in the beginning, to reflect on my feelings of mathematics on paper, since I had never been asked to do such a task before."

A reflection tool

Pre-service teachers also noticed that writing allowed them to step back and reflect on what they did. A student wrote about one of the writing assignments:

"The most important item... was the Standards for Teaching of Mathematics. The activity, which we had to write on, helped me clarify my own beliefs about mathematics teaching and how purposeful simple things can be. Up until writing about the strategies, I had not realized the obvious things we were doing in math class and how it directly applied to our own classrooms, for example, cooperative learning.... This was the most important item to me because it helps me see the full picture instead of looking at bits and pieces."

Writing also offered prospective teachers the opportunity to reflect on how children learn mathematics. One teacher wrote about his own experience: "I found it refreshing to learn about fractions in a concrete fashion. This is how children learn best, and sometimes to learn a concept we have to become as little children." Another teacher reflected on children's learning:

"My favorite item [...] would be my first teaching lesson. In it, I had the children use manipulatives to complete an addition worksheet. I learned so much just from watching my children. I was fascinated by how they chose to do the addition. Some needed the blocks to help them see the addition while some did not. Some had to add similar colors while others did not seem to care. I probably learned more about how children think and think so differently from others in that short amount of time than ever before. How grateful I am for that experience of seeing through the eyes of a child." Writing allowed this teacher to keep a record of such a memorable and crucial experience in her teaching career. Another teacher wrote:

"I learned… to let students learn and explore for themselves. I learned to help them by giving them the room they need to discover things for themselves."

Taking the pressure off: Supporting a positive learning environment via writing
Writing activities can also support a positive learning environment. In many cases, the pre-service teachers experienced, when they were students, the pressure to come up with the *right answer* using the teacher's methods, which made the learning of mathematics a very unpleasant experience for some of them. Writing in the mathematics methods course helped students release the anxiety. By writing about their experiences with mathematics, these pre-service teachers were expected to organize their thoughts in a more logical sequence and create meaning from their experience, even if they did not come up with the most efficient strategy. Some teachers' reactions to writing in the mathematics methods course express how it helped them to eliminate the anxiety about math:

"Usually, after we finish all the activities for the day, the professor had us write a reflection on our learning. I liked this because it allowed me to make further sense of my meaning making, as I write my feelings, thoughts, and emotions on paper."

Another teacher expressed, "To be able to write down what I really felt about mathematics was very cathartic for me."

Writing in the mathematics methods course from an *affective approach* had two benefits. The first was that writing in the classroom changed the expectations of the pre-service teachers as students—from answer-producers to critical thinkers. They were expected to reflect on their feelings, beliefs, and thinking processes about mathematics. The same could happen with their future students if these pre-service teachers choose to use writing.

The second benefit of writing in the mathematics methods course was helping to reflect on a cooperative learning environment. For example, during the course, the future teachers expressed that working in cooperative groups had many positive effects in their own learning. By sharing and discussing their writing with others, they were able to learn from each other and expand their repertoire of strategies to pursue mathematical processes. However, a good group relationship does not happen automatically and this was reflected in some teachers' reactions:

"When we worked in cooperative groups I felt intimidated by other individuals in my group. The way I went about solving the problems were different than others. I needed the more visual and manipulative capabilities accessible to me, whereas my group attempted to use formulas and more abstract methods."

These comments remind us of the importance of the teacher to facilitate a good communication and cooperation among the groups. Cooperative learning does not take place by just putting a group of people to work together on a task. It is important to make sure that all ideas are valued and are validated by all members of the group. Again, students may use writing to share their concerns.

Writing for an audience

By providing children with opportunities to use speaking, reading, and writing for authentic purposes, children learn to value the functional uses of language skills in their lives. Likewise, pre-service teachers became involved in the writing tasks because they found a functional or authentic purpose for the writing: All the writing activities had the potential to be implemented in a real classroom. But most importantly, the writing assignments in the mathematics methods course had the purpose to establish a dialogue between the pre-service teachers and the professor to expose ideas, concerns, and suggestions about how the course was fulfilling the needs of each pre-service teacher. The written dialogue between professor and future teachers served the authentic purpose of establishing a genuine communication to help the professor adjust the course to their needs. Writing in the mathematics methods course allowed these pre-service students to share their own experiences, positions, and attitudes in mathematics learning and receive feedback.

One important element of the use of writing in the methods course is that the professor is perceived as someone more interested in the communication, rather than assessment (Johnson et al, 1993). By responding to all the writing assignments, the professor helps future teachers to understand this role. The professor responding to their writing was something that students noticed in this course. One student wrote in her final portfolio about one of the early reactions:

"I was still very unaccustomed to writing about math [...]. I had just begun to realize that our professor indeed read everything we wrote and always replied. There was a dialogue. Again, something I had not experienced."

When the professor writes comments back to his/her students, he/she participates in the dialogue. Then, the instructor becomes part of the audience:

"I find these reflections beneficial because it gives me a way to let you [the professor] know what point I am at and it also is my way of writing down what I thought helped me and what I thought didn't quite work. I think the reflections are great because it is a way that I know that I can show you that I am either confused or progressing in math."

Prospective teachers also write to ask questions about points in the chapter or in class that were not clear, and some of the questions required a lengthy response.

Some teachers found beneficial to share their writing with their peers. The importance of sharing is that it allows student to negotiate meaning and expand their knowledge. Part of the drafting stage of the writing process involves "sharing" for negotiation of meaning, making sure that the audience understands the message as intended (Peterson, 1994). Sharing might also provide an opportunity to learn from peers who may have a different perspective about a common concept or learning situation. In this course, teachers had the opportunity to work in pairs and small groups. Each member of the group wrote a short strategy about how to solve a teaching problem. Some students found that writing forced them to think on how to communicate their ideas to others. Then the members shared their approaches to engage in a group discussion. When students interacted to do the activity and to negotiate what would be reported, sharing of ideas and clarification took place. This sharing allowed the teachers to see their peers' perspectives. As one pre-service teacher explained: "Today, I realized that the way that seems so obvious to me is not necessarily the obvious one to others." After the discussion, a general report was produced. Sometimes, all the approaches were included and sometimes the teams came up to a consensus and just one approach was selected.

Another benefit that some teachers expressed was that writing allowed them to learn about themselves and their growth and changes over the semester. The written record let them contrast their positions at different points in time. At the beginning of the semester, one of the students wrote in her biography: "math is one of the few subjects which is absolute -- the answer is right or wrong and the feedback is immediate." At the end of the semester, she reflected how her perception had changed:

"My training in mathematics as an elementary and high school student geared me toward this kind of thinking. Since this class, I think that mathematics includes different ways of thinking about math problems and ways to solve problems. I can see that not all people look at math problems the same and that different perspectives come into play."

Another teacher reflected on how the meaning of *understanding mathematics* had changed for her, from the first session to the end of the semester:

"I said that understanding math comes fairly easy to me; however, after exploring concepts relationally, I know that I never really understood math at all. I simply could memorize formulas and plug numbers into them with no idea of why the formula worked. I had only instrumental understanding of math."

Change is not easy. Part of the difficulty is that one semester barely seems enough:

"I feel like even though I have a solid foundation of mathematical concepts, I am continually reconstructing the way that these are to be taught. The difficulty is that the way I am being trained to teach is a way that I have been taught only for one semester."

Skemp (1987) recognized "the great psychological difficulty for teachers of reconstructing their existing and longstanding schemas" (p. 161). Writing was a way to document the difficulties encountered by these pre-service teachers in the road to change their way of thinking about mathematics. Pre-service teachers could look back at their writing and reflect on their resistance to change:

"In the beginning, it was difficult for me to go back to concrete concepts and explain them.... [It] was hard for me to look away from what I was taught in mathematics and open my eyes to a better way."

CONCLUSION

In this paper, we have discussed how writing was applied in a mathematics methods course as a reflective tool for pre-service teachers. The accounts of these students in the mathematics methods course revealed four major benefits of writing:

• Writing as a tool to elicit background knowledge and framing mathematics into a context that is personal and relevant to the students;

• Writing as a tool for students to organize and put in perspective their own thinking about teaching mathematics;

• Writing as a reflection tool to help students to have a permanent record of the evolution of their ideas and learning about the teaching of mathematics for future reflection and enhanced understanding; and

• Writing to support cooperation and dialogue among students and teachers for mutual learning.

One of the activities that best captured the attitudes of pre-service teachers towards writing was the building of portfolio. At the end of the semester, students selected five items from the work they produced during the course to create a portfolio. The goal was to include items that showed *growth* in their understanding of how to teach mathematics. The process of reading and collecting their writings allowed them to look back at the semester as a whole and realize how they changed:

"As I went through all of my work to select the five items to be included in my portfolio [...] I found myself really enjoying the reflection on where I started and where I've come. It was interesting to review my work, feelings, thoughts, and beliefs as they developed, grew, and changed over the semester. I was surprised by a few things, mainly how my belief and attitude toward teaching math was changed. I'm going to enjoy the challenge now and I feel much better knowing that it's more important to find out what's meaningful to and for the student than to push skill, drill and rote paper and formula exercises."

For the scope of the paper, we have tried to suggest the use of writing in the mathematics class as a tool, as a complement to other mathematics methods and lessons. Some pre-service teachers expressed some of the benefits of incorporating writing activities along with other tasks and assignments,
"A very integral part of the course for me were the reflections/reactions and creative activities. I had not before ever written out my thoughts/reflections in regards to mathematics and found this experience vital to my learning experience."
Another one stated,
"When the class began and we were asked to reflect on different things, readings or class activities, I was very uncomfortable about writing about math. I feel where I showed the most growth is in these reflections, in thinking about what math means, how to teach it, and how to make it meaningful for the children. These reflections changed my whole attitude and belief of math instruction. It was very enlightening activity, one I never imagined would occur in a mathematics class."

Teachers in this mathematics methods course used writing as a tool to reflect on their own learning and growth as pre-service teachers of mathematics. We believe that the incorporation of writing assignments might add a new element in the preparation of teachers of mathematics. As a supportive tool to other methods of teaching mathematics, writing to reflect helped these prospective teachers to contextualize, organize and reflect on their mathematical concepts and the way they intended to teach mathematics in the future.

REFERENCES

Connolly, P. and Vilardi, T. (Eds.). (1989). Writing to learn mathematics and science. New York, NY: Teachers College Press.

Countryman, J. (1992). Writing to learn mathematics. Portsmouth, NH: Heinemann.

Drake, B. M. and Amspaugh, L. B. (1994). What writing reveals in mathematics. Focus on Learning Problems in Mathematics, 16 (3), 43-50.

Freeman, Y. S. and Freeman, D. E. (1992). Whole language for second language learners. Portsmouth, NH: Heinemann.

Gribbing, W. G. (1991). Writing across the curriculum: Assignments and evaluation. The Clearinghouse, 64, 365-368.

Johnson, J., Holcombe, M., Simms, G., and Wilson, D. (1993). Writing to learn in a content area. The Clearinghouse, 66(3), 155-156.

Kneeshaw, S. (1992). KISSing in the history classroom: Simple writing activities that work. The Social Studies, 83(4), 176-179?

McIntosh, M. (1991). No time for writing in your class? Mathematics Teacher, 84(6), p. 423-433.

Mett, C. L. (1989). Writing in mathematics: Evidence of learning through writing. The Clearinghouse, 62, 293-296.

Peregoy, S. F. and Boyle, O. F. (1993). Reading, writing, and learning in ESL: A resource book for the K-8 teachers. White Plains, NY: Longman.

Peterson, R. (1994). Grand conversations: Literature groups in action. New York: Scholastic Books, Inc.

Skemp, R. R. (1987). The psychology of learning mathematics. Hillsdale, NJ: Lawrence Erlbaum Associates.

12 HOW MANY TIMES DOES A RADIUS SQUARE FIT INTO THE CIRCLE?[12]

Abstract

The first part of the article illustrates a way in which students can estimate the ratio of the area of the circle to the radius square. The second part helps students understand why the same value π appears in both the formulas for the circumference and the area of the circle.

In this article we present alternative and interesting approaches to explore π. In the first part, we describe an empirical method that allows students to estimate the area of a circle with surprising accuracy. It can be used as an extension of the approach of using a grid to approximate the area of a circle. The use of a grid is conceptually very illuminating; it emphasizes that the area is measured in square units, and that areas of shapes that do not have straight sides can nevertheless be measured with square units. An example of such an approach is *Covering a Circle* (Lappan, Fey, Fitzgerald, Friel, and Phillips, 1998). However, when students actually count squares, their values may not be very exact. For example, in a sixth grade classroom, when counting the squares in a circle with radius 7 contained in a square grid, students' answers ranged from 142 to 178 square units. Dividing such numbers by the square of the radius, that is 49, we obtain values ranging from 2.9 to 3.6. There was not enough exactness to determine the value of π with two significant digits. In the first part we want to help students to understand that the area of the circle is about 3.1 times the area of the radius square.

Frequently, after allowing students to explore empirically both the area of the circle and its circumference, they are not helped to understand why the same number π appears in both the formulas for the area and the circumference of a circle. The second part of the article is geared to help students see the connection. This may be more appropriate for a re-visit of the topic in seventh or eighth grade. We assume that students have done an activity for finding the ratio of the circumference to the diameter, for example, by measuring around circular objects, measuring along the corresponding diameters, and computing the ratio. An example of such an activity is *Surrounding a Circle* (Lappan, Fey, Fitzgerald, Friel, and Phillips, 1998). Also, we assume that students know and understand the formulas for the area of a triangle and a parallelogram.

Part 1. The ratio of the area of the circle and the radius square.
Students are given a circle and a square whose side is equal to the radius, that is, the radius square (fig. 1). They are asked to guess how many times the radius square fits into the circle. Students in sixth grade figured out fairly quickly that the answer had to be more than two but less than four. Figure 2 shows a way by which students may realize that the radius square fits more than two times in the circle. The area of the square inscribed in the circle is equal to two radius squares, because the area of the shaded part is one half of the radius square, and there are four such halves. Figure 3 shows that the radius square fits less than four times because four radius squares cover and extend beyond the circle.

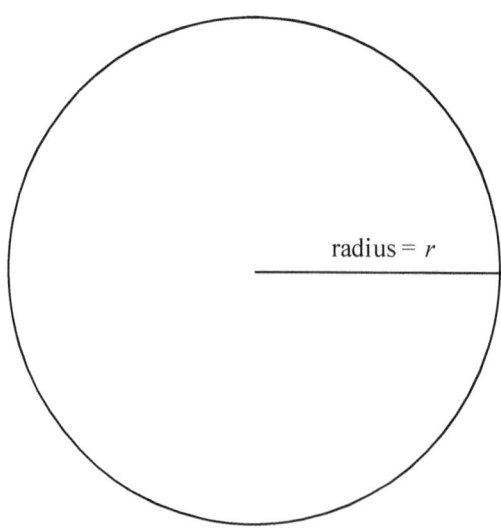

radius = r

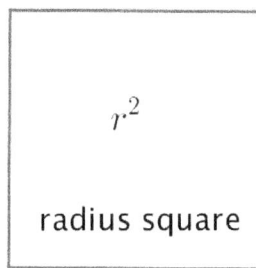

Figure 1. The circle and the radius square.

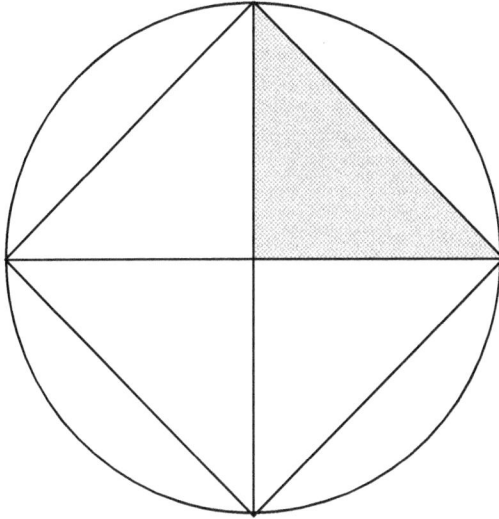

Figure 2. The shaded triangle is one half of the radius square.

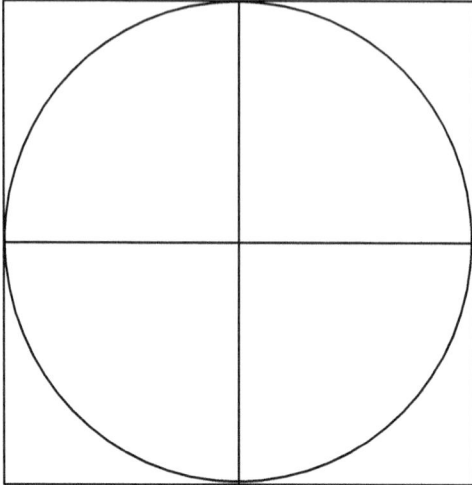

Figure 3. The circle is less than four radius squares.

<u>Activity 1. Cutting and pasting radius squares.</u>
<u>Instructions for students</u>. Color the four radius squares (at the end of the article) with different colors. Cut out the circle. Cut out the first square and fit it entirely inside the circle. Cut out the second square and fit it inside without overlapping the previous color. You will have to cut parts of the second square so that they fit inside the circle without extending beyond it. Figure 4 shows one possible response; the pieces of the squares fit in other ways. Use all of the second square before using the third square. Continue with the third and then the fourth squares in the same manner. Save the remainder of the fourth square and use the grid to estimate how much of the fourth square you were able to fit.

Students see that they can fit three squares completely and a little of the fourth one. That is, the radius square fits "three and a little more" times into the circle. It is convenient to use a radius of ten units and the corresponding ten by ten grid to draw the fourth radius square. By counting how many of the small unit squares of the fourth radius square were actually used, students can see that the ratio of the areas of the circle and the radius square is about 3.1.

The activity took about two 45-minute sessions to complete in a sixth grade class. The teacher may want to have wax paper, patty paper, or other translucent paper available so that students can trace areas that are not yet covered on the circle and use the trace to cut the corresponding parts on the colored radius squares. Students have different strategies to fit parts of the fourth square, some of which lead to better estimates. However, in the sixth grade class mentioned, students' estimates of the ratio of the area of the circle to the radius square were all between 3.10 and 3.20.

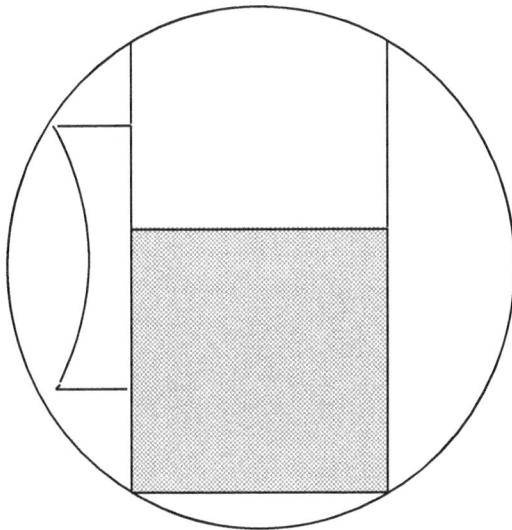

Figure 4. Student fits two radius squares into the circle before using the third one.

By using circles of other sizes, students can also be convinced that the ratio between the area of the circle and the radius square is the same for circles of any size. At this point some students will notice that this value of 3.1 is the same as the ratio of the circumference to the diameter and may guess that it is not a coincidence. The next part helps students see why the same constant π appears in both the formulas for the circumference and the area of the circle.

Part 2. The relation to π in the formula of the circumference.
Students need to recall that the ratio of circumference to diameter is approximately 3.1. The exact value of the ratio is called π, and its first digits are 3.14, which for most practical purposes gives enough accuracy. Many calculators have a key for π in case more digits are needed.

<u>Activity 2. Approximating the circle by regular polygons. (A thought experiment.)</u>
Imagine you have a family of regular polygons inscribed in the same circle constructed in the following way. Starting with a regular hexagon (figure 5a),

the next polygon will have 12 sides. Six of the vertices will be common with the hexagon; the additional vertices will be the midpoints of the arcs (see figure 5b). In the same way, each successive term of the family of polygons has twice as many sides. The mores sides in the regular polygon, the closer the perimeter of the polygons is to the circumference of the circle. Furthermore, by using a polygon with a large enough number of sides, we can make the difference between the perimeter of the polygon and the circumference as small as we want. The areas of the regular polygons approximate better and better the area of the circle. The difference between the area of the circle and the area of one of the polygons can also be made as small as we want by choosing a polygon with a sufficiently large number of sides.

The area of the regular polygon can be computed by multiplying the perimeter times the height of one of the triangles forming the regular polygon (see figure 6), and dividing by two. This can be proved in several ways. One is to imagine all the triangles laid out side by side (figure 7). The total area of the polygon is the sum of the areas of the triangles. One method to obtain the total area is to compute the area of each triangle by multiplying the base times the height, divide by 2, and then add the areas. Alternatively, we can add all the bases first, which gives us the perimeter, then multiply by the height, and divide by 2. As the number of sides on the polygon increases, the sum of the bases will be very close to the circumference of the circle ($2\pi r$), and the height of the triangle will be very close to the radius (r). Therefore the area of the polygon will be very close to $\dfrac{circumference \times radius}{2}$. Because the area of the circle and that of the polygons can be made as close to each other as we want, we have the area of the circle given by

$$\frac{circumference \times radius}{2} = \frac{diameter \times \pi \times radius}{2} = \frac{2 \times radius \times \pi \times radius}{2} = \pi \times radius^2$$

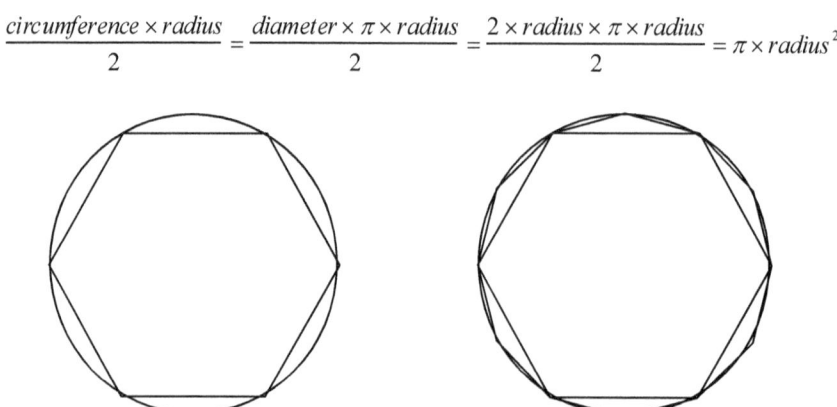

(a) The initial hexagon   (b) New vertices at the midpoints of the arcs
Figure 5. Regular polygons that approximate a circle.

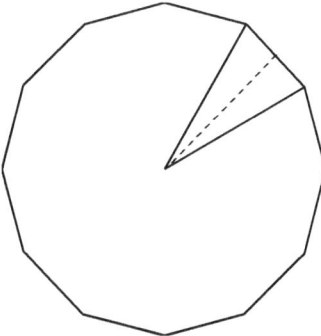

Figure 6. A triangle whose base is one side of the polygon.

Figure 7. The polygon is broken into triangles.

We can also arrange the triangles that form the regular polygon into a parallelogram (see figure 8). Its base will be very close to half the circumference, $\frac{1}{2} \times d \times \pi = \frac{1}{2} \times 2 \times r \times \pi$, that is $r \times \pi$, and its height will be very close to the radius r of the circle. The area of the parallelogram will therefore be very close to $\pi r2$. As the number of sides of the regular polygon increases, the height of the corresponding parallelogram gets closer and closer to the radius of the circle, and its base closer to πr. The area of the circle is given by $\pi r \, \square \, r = \pi r2$.

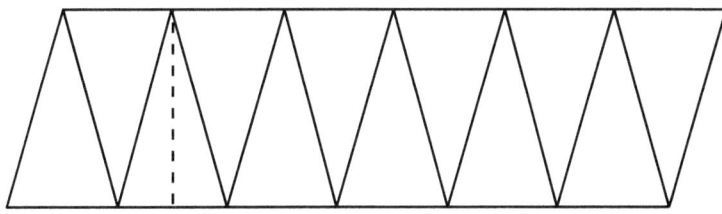

Figure 8. The polygon is rearranged into a parallelogram.

Activity 3. A different approach.

Figure 9 shows a circle of radius 5 cm. Its circumference can be computed by using the formula $2 \times r \times \pi$; in this case the circumference of the circle is $2 \times 5 \times 3.14 = 31.4$ cm. Imagine that you cut out the circle into 16 sections and rearrange them to form a shape that resembles a parallelogram as shown in figure 10. The area of the circle is the same as the area of this shape. To compute the area of this shape we will use the fact that it resembles a parallelogram, although its "base" is not quite a straight line. We will use the formula for the area of a parallelogram, Area = base × height. The length of the "base" of this "parallelogram" is half the circumference of the original circle, because half of the sections point down, and the other half point up. So the length of the base is close to 5π (if you measure along the base on a straight line you will find the length is about 15.7 cm). The height of the "parallelogram" is very close to the length of the radius of the original circle, that is 5. The area of this "parallelogram" will be close to the length of the "base" times the height, that is $5 \times 5\pi = 25 \pi$. Notice that 25 is the value of the radius squared. Because the area of the "parallelogram" is equal to the area of the circle, we can see that this allows us to compute the area of the circle by squaring the radius and multiplying by π. As you see, the ratio of the area of the circle to the area of the radius squared is precisely π. The fact that in our first activity we obtained 3.1 for the ratio of the area of the circle to the radius square, the same value as the ratio of the circumference to the diameter, was not just a coincidence.

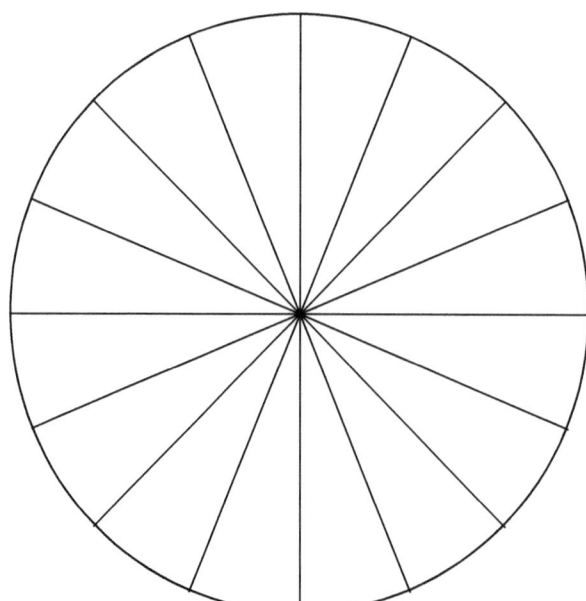

Figure 9. A circle cut into 16 slices.

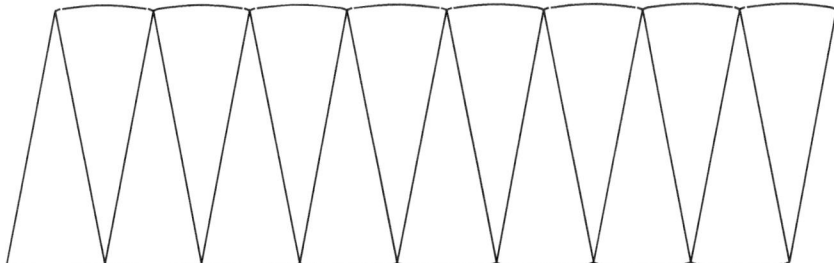

Figure 10. The slices of the circle rearranged.

Instead of cutting the circle into 16 parts, suppose we cut it into more sections, say 36, and rearrange the sections as before, the new shape will resemble a parallelogram even more (figure 11). The length of the "base" of this new "parallelogram" is half the circumference, that is $5 \times \pi$, and it is even closer to being a straight line, and the height of the "parallelogram" is even closer to the radius, that is 5. We can conclude that the area of the circle is given by 25π, that is $5^2\pi$.

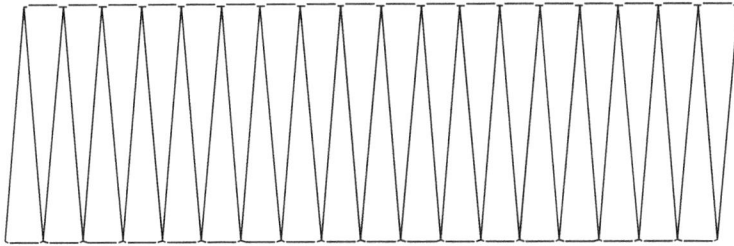

Figure 11. A better approximation to a parallelogram of height r.

We can imagine the same process with circles of a different radius. If the radius of the circle is 4, the circumference would be $2 \times 4 \times \pi$; half the circumference would be $4 \times \pi$. This would be approximately the "base" of the "parallelogram". The height of the "parallelogram" would be approximately 4, so its area would be $4 \times 4 \times \pi$, or $4^2 \pi$. In general, if the circle has a radius of length r, its circumference will be $2r\pi$. Half the circumference will be $r\pi$. So, if we cut the circle into thin slices and rearrange them to form a "parallelogram," the length of its "base" will be $r\pi$ and its height will be very close to r. Its area will be therefore be very close to $r^2\pi$. So, the ratio of the area of the circle to the radius squared $\pi r^2 / r^2$ is precisely π.

The two methods discussed in the second part of this article are closely related (see figure 12). Some students may prefer working with real parallelograms that approximate better and better the area of the circle. Others may prefer working with families of shapes, all of which have the same area as the circle, and that resemble more and more a real parallelogram.

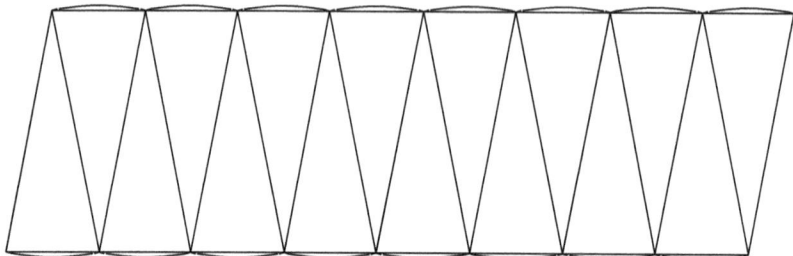

Figure 12. A real parallelogram and the circle rearranged.

Concluding remarks

According to Van Hiele (1999) students progress through levels of geometrical thinking. Learning opportunities and guidance from the teacher will help them make the transition from one level to the next. In the middle grades, a particularly important transition is from the level of development where students rely heavily on empirical verification to a level where deductive and more abstract thinking plays an increasingly important role. It is important that the approach to mathematics, while still based on concrete and visual representations, gradually relies less on empirical measurement and more on thought experiments and convincing arguments. By allowing students to approach the topic of the area of the circle, first through an empirical approach and later through a more deductive one, we hope to help students in the middle grades make that important transition.

References

Lappan, Glenda, James T. Fey, William M. Fitzgerald, Susan N. Friel, and Elizabeth D. Phillips. *Covering and surrounding*. Connected Mathematics - Geometry. Menlo Park, CA: Dale Seymour Publications, 1998.

Van Hiele, Pierre M. "Developing Geometric Thinking through Activities that Begin with Play." *Teaching Children Mathematics*, 5 (February 1999): 310-316.

Materials

The squares provided are radius squares of the given circle. Color each square with a different color. Cut the squares and cut out the circle. Follow the instructions for Activity 1.

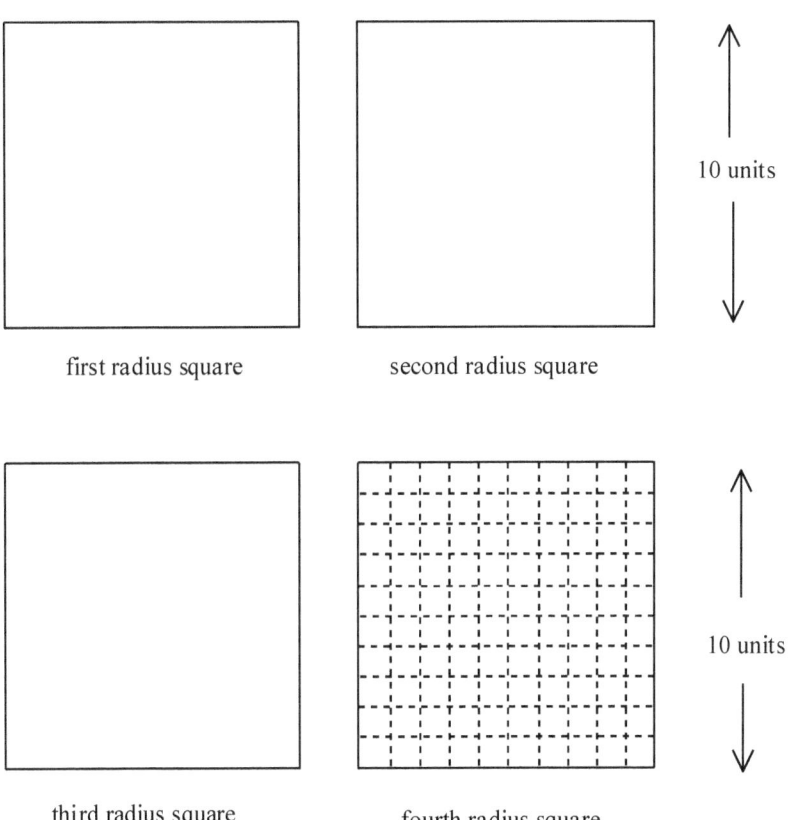

first radius square

second radius square

10 units

third radius square

fourth radius square

10 units

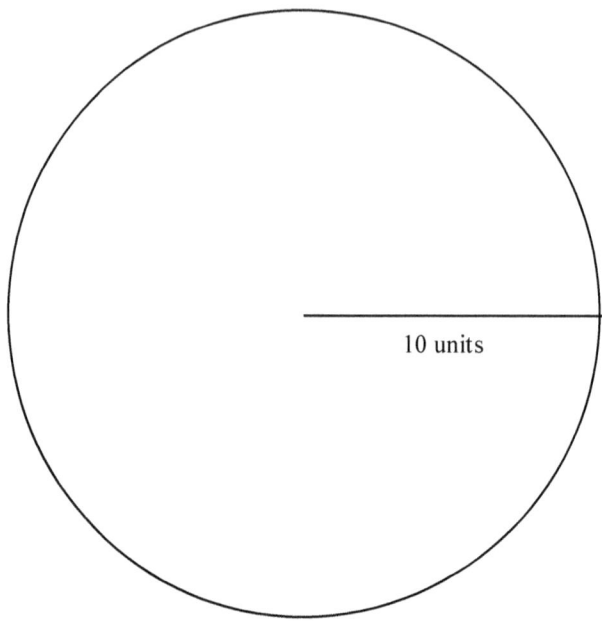

10 units

13 GEOMETRIC REPRESENTATIONS IN THE TRANSITION FROM ARITHMETIC TO ALGEBRA[13]

Prelude. Some striking numerical results (the reader may want to use a calculator).

1) Take the square of a whole number, add the number, and add the next number. The result is the square of the next number.

$$5^2 + 5 + 6 = 6^2$$
$$10^2 + 10 + 11 = 11^2$$

2) Take any odd number, square it, and subtract one. The result is divisible by 8.

$$7^2 - 1 = 8 \times 6$$
$$9^2 - 1 = 8 \times 10$$

3) Consider the following triangle of odd numbers. Add the numbers in each row. The result in each case is a cubic number.

1	1	1^3
3 + 5	8	2^3
7 + 9 + 11	27	3^3
13 + 15 + 17 + 19	64	4^3

4) Take a number, raise it to the third power, and subtract the number. The result is divisible by six.

$$3^3 - 3 = 24 = 6 \times 4$$
$$4^3 - 4 = 60 = 6 \times 10$$
$$5^3 - 5 = 120 = 6 \times 20$$

5) Multiply four consecutive numbers. The result differs by one from a

---

[13] Flores Peñafiel, A. (2002). Geometric Representations in the Transition from Arithmetic to Algebra. In F. Hitt (Ed.), *Representations and Mathematics Visualization* (p. 9-29). México: Departamento de Matemática Educativa del CINVESTAV-IPN.

perfect square.

$$1 \times 2 \times 3 \times 4 + 1 = 5^2$$
$$2 \times 3 \times 4 \times 5 + 1 = 11^2$$

Introduction

The continuities and discontinuities between arithmetic and algebra that students face are complex and need to be addressed from multiple perspectives. There are several aspects that teachers need to take into account in order to help students make the transition from arithmetic to algebra, and to develop algebraic skills with understanding (see for example Kieran & Chalouh, 1993; Wagner & Parker, 1993, Lodholz, 1990). This chapter focuses on one possible way to help students develop their algebraic thinking, using geometrical representations of interesting numerical relationships. The numerical examples provided are of the kind that students find striking or amazing (Mulligan, 1988). Using geometric representations for these numerical relations can help students go from statements about particular numbers to the corresponding generalized statements using variables. These representations provide a way to shift students' attention, from the purely procedural approach to numbers, to considering the terms and operations involved in a numerical relationship as entities that are worthwhile to pay attention to. Geometrical representations can provide a context where students

- learn to extract pertinent relation from problem situations and express those relations using algebraic symbols;
- make explicit the procedures they use in solving arithmetic problems;
- consider strings of numbers and operations as mathematical objects, rather than processes to arrive at an answer;
- gain explicit awareness of the mathematical method that is being symbolized by the use of both numbers and letters;
- focus on method or process instead of on the answer;
- pose and compose problems;
- write conjectures, predictions, and conclusions.

Geometric representations of relationships among numbers go back to antiquity. For example, Euclid, in book II includes the following statement and diagram (fig. 1), "If a straight line be cut at random, the square on the whole is equal to the squares on the segments and twice the rectangle contained by the segments" (Euclid, 1956, p. 379).

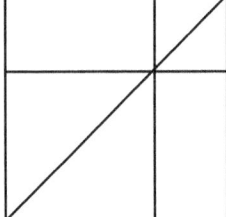

Figure 1. Squares on a segment

Some students also spontaneously use geometric representations to convince themselves of the truth of a statement. Krutetskii (1976, p. 325) gives the example of a sixth grade student who drew a figure to convince herself and better understand the formula for the square of a sum of numbers $(a + b)2 = a^2 + b^2 + 2ab$.

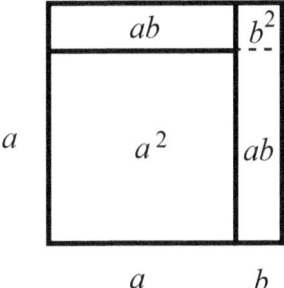

Figure 2. $(a + b)^2 = a^2 + b^2 + 2ab$.

The importance of the use of geometrical representations for algebra has been recently reiterated (Algebra Working Group, 1998; Bass, 1998). One of the components of the geometry standard (National Council of Teachers of Mathematics, 2000) emphasizes that instruction should enable students to use visualization, spatial reasoning, and geometric modeling to solve problems. For the grade band 6-8, it recommends that students "use geometric models to represent and explain numerical and algebraic relationships." (p. 232). Arcavi (1999) recommends to "bring geometry to the aid of what seem to be purely symbolic / algebraic properties." (p. 60)

These recommendations are supported by research on how humans learn. According to Fischbein (1987), visualization "is the main factor contributing to the production of the effect of immediacy" (p. 103). In mathematics as in many other fields, "the concreteness of visual images is an essential factor for creating the feeling of self-evidence and immediacy" (Fischbein, 1987, p.

104). Presmeg (1999) found that "visual methods were effective when they were used in ways that supported generalization" (p. 152). Botsmanova (1972a) explains why the use of pictorial visual aids helps to bridge the gap between the concrete situation reflected in the problem's subject and its abstract side—the mathematical structure. According to her, "any pictorial representation combines features of the abstract and the concrete. A pictorial representation is the abstracted and generalized expression of a rule, and at the same time it translates the solution of a problem from the abstract-verbal level into a concrete plan." (p. 105). Furthermore, Botsmanova (1972b) states that the interlacing of the abstract and the concrete in a graphic expression is clearly what permits the use of a graphic diagram in problem solving both in the transition from abstract to concrete and the reverse transition, from the concrete or visual to the abstract. The ability to operate with a diagram can permit a pupil to discover mathematical relationships more easily, and can become, under certain conditions, a generalized method of analysis and synthesis. (p. 119)

Geometrical representations can help focus the attention of the students on the relations in general and help them use letters to stand for given quantities; use algebra as a tool for proving rules governing numerical relations; use the equal sign to express a symmetric and transitive relation; use letters to represent givens in the generalization of number patterns; and use letters to specify a range of values in representing numerical relationships.

Geometrical representations of numeric relations can be a means to explore algebraic ideas. With them students can think about the relations, discuss them explicitly using ordinary language, and learn to represent them with letters. These representations can serve as a guide for students as they learn to use algebra as a tool for generalization and justification. Geometrical representations can serve as a scaffolding as students develop proficiency with algebra as a language for actions on quantities and relationships among quantities. They can provide a semi-concrete step towards the more abstract concepts that arise with the use of letters as variables. They can help students avoid over generalizing and help students not to judge algebraic expressions based on superficial characteristics. They can help students to develop a more complete mental construct of variable by considering a more ample range of cases of the concept of variable, by using them to represent given quantities and not just as abbreviations or unknowns.

Geometrical representations can help students see that multiplying and factoring are inverse operations by using area or volume models of multiplication. They can also serve to illustrate cases when more than one letter in a parenthesis are dealt with together, as a single entity, a necessary ability for algebra which is sometimes hampered by experiencing only procedural ways to deal with quantities in parenthesis, such as "do what is in parenthesis first" or "clear the parenthesis first." As Warner and Parker

(1993) point out, the use of pictorial representations together with concrete numerical cases should help students apprehend the equivalence of algebraic identities in symbolic form. Visual models and numerical examples can reinforce structural relationships in algebra.

Kieran (1990) states that "the power of the symbolic language is that it removes many of the distinctions that the vernacular preserves, thus vastly expanding its applicability. The cost is that the symbolic language is semantically extremely weak" (p. 97). Geometrical representations can provide in some cases a semantic support for dealing with the symbols.

The purpose of the examples that follow is to give some ideas for teachers how to help students make the jump from the visual and concrete to the symbolic and abstract. In algebra and beyond, students need the ability to deal with symbols for variables rather than just symbols for numbers. In addition, algebraic notation will help them reason about statements that apply to all numbers or to numbers in a specified set, rather than about statements that apply to particular numbers.

Teachers in the middle grades or in the first year of high school can help students develop their ability to deal with symbols, as students use concrete representations of relations among ordinary numbers to serve as a bridge to symbolic notation. As students write equations representing several cases and compare them, they will see the pattern and will be able to express the equations using variables.

An important step in understanding is to describe what the figures and the equations represent using plain language. Verbalizing the steps and using drawings or other concrete representations can also help students make the connection with the symbols. It is important to see how parts and terms of the equations are represented by parts of the figures. The drawings and particular numbers used will help students attach meanings to the different terms in the algebraic formula used to describe the general situation.

Although the figure itself only illustrates a particular case, the figure contains the elements that will facilitate general reasoning.

Of course, students can benefit from the use geometric representations years before the transition to algebra. For example, an area model can make explicit the use of the distributive property in a common procedure to multiply two-digit numbers (figure 3).

$$
\begin{array}{rr}
23 & 20 + 3 \\
\times\ 12 & \times\ 10 + 2 \\
\hline
46 & 40 + 6 \\
230 & 200 + 30 \\
\hline
276 & 200 + 70 + 6
\end{array}
$$

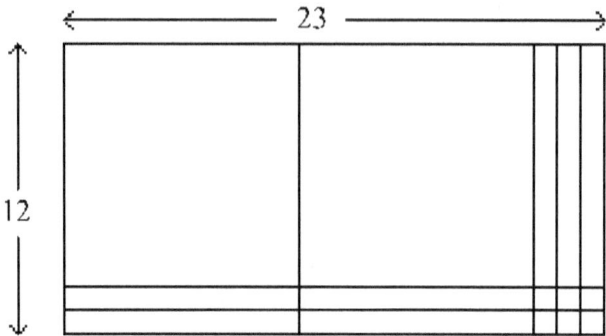

Figure 3. Area model for 12 × 23

In the following sections we will provide some examples that can be used in the classroom in the middle grades. For more examples of geometric representations of interesting numerical or algebraic relations, the reader can consult Nelsen (1993), as well as previous published work by the author (Flores 1992; Flores Peñafiel 2000a, 2000b, 1999).

Examples for the classroom

Example 1. Consecutive squares

Students can see why starting with the square of a number, and then adding the number, and the next number does always give the square of the next number by using a drawing (figure 4). They can realize that the drawing, although made for a particular number, permits reasoning that is general and does not use special properties of the particular numbers (4 and 5).

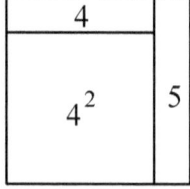

Figure 4. $4^2 + 4 + 5 = 5^2$

Using this as their starting point they may later use a more general diagram (figure 5). Of course, once students have developed skills in algebraic manipulation the equation can be easily verified. The point, however is that the geometric representation can give meaning to the different terms of the equation, $n2 + n + (n + 1) = (n + 1)2$ and students can use it to make sure the symbol manipulation is correct.

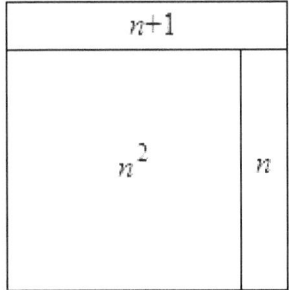

Figure 5. $n^2 + n + (n + 1) = (n + 1)^2$

Example 2. Sums of consecutive numbers.

The sum of consecutive numbers $1 + 2 + 3 + 4$ can be represented by columns of squares of increasing heights (figure 6). Adding another number would mean adding another column to the staircase. If n numbers are added, then the highest column would have n squares. The sums of consecutive numbers are called triangular numbers. Thus $10 = 1 + 2 + 3 + 4$ is the fourth triangular number.

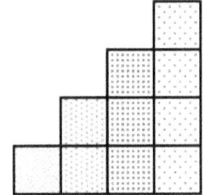

Figure 6. $1 + 2 + 3 + 4$

Students can make another exact copy of their stair, and use the two copies to form a rectangle (figure 7). In this case it is a 4 by 5 rectangle, but students can realize that in general it will be a rectangle in which the lengths of the sides are two consecutive numbers. The shorter side will indicate how many numbers are being added. If they add the first n numbers, it will be an $n \times (n+1)$ rectangle.

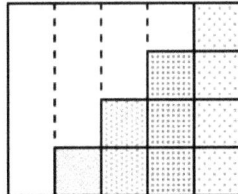

Figure 7. A 4 by 5 rectangle

Because the rectangle has two copies of the triangular number, we need to divide its area by 2. Therefore, students can see that $1 + 2 + ... + n = \frac{n(n+1)}{2}$. The expression $\frac{n(n+1)}{2}$ can be algebraically transformed into $\frac{n^2}{2} + \frac{n}{2}$. Students can attach meaning to the terms of this expression using a big triangle and n half squares (figure 8).

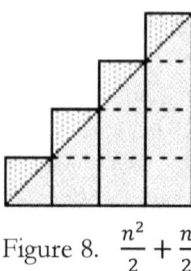

Figure 8. $\frac{n^2}{2} + \frac{n}{2}$

Example 3. The square of an odd number minus one.
Ask students to take any odd number, square it, subtract one, and then divide the result by 8. They will be surprised to see that for all the odd numbers they chose, the division does not leave a remainder. Students can try different odd numbers in a systematic way, and organize data as in table 1. This will provide quite some empirical evidence that the result is true for any odd numbers. However, empirical evidence is not enough in mathematics; students need to understand why it works. Some students may recognize the triangular numbers in the last column of table 1. Based on this observation they may predict what the next result will be.

Table 1. Squares of odd numbers, minus one

Odd number	squared	minus one	divide by eight
1	1	0	0
3	9	8	1
5	25	24	3
7	49	48	6
9	81	80	10
11	121	120	15

A geometric representation can help students understand why when we subtract one from the square of an odd number the result is always divisible by eight Students are given a square (with a side of odd length) that is missing a unit square (figure 9a). In general, because odd numbers are of the form $2n+1$, we can describe the square minus one unit as $(2n+1)^2 - 1$.

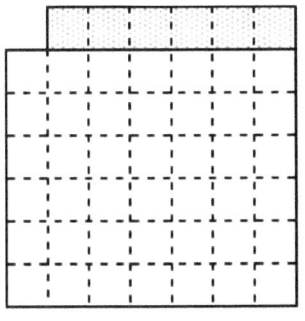
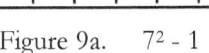
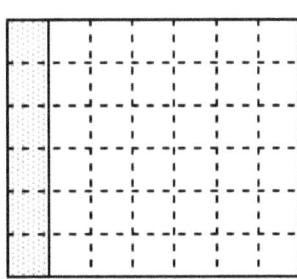

Figure 9a. $7^2 - 1$ Figure 9b. 6×8

By rearranging the colored strip, students can form a rectangle with the same area. Because the original number is odd, both sides of the rectangle will be even numbers, one side two units longer that the other (figure 9b). In general, we can describe this rectangle as $2n \times (2n+2)$. Therefore, each side can be divided into two equal parts and giving rise to four rectangles (figure 10a); the length of each side of the smaller rectangle will be a whole number. We can express the area of the big rectangle as $4 \times n \times (n + 1)$. The sides of the small rectangles differ by one unit. That is, they are the kind of rectangles we obtained by putting two triangular numbers together. Students can see that eight triangular numbers are hidden in the rectangle, $8 \times \frac{n(n+1)}{2}$ (figure 10b)

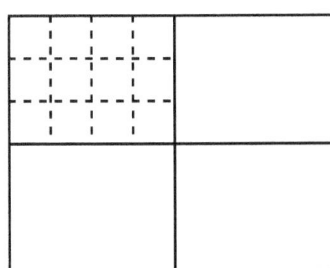
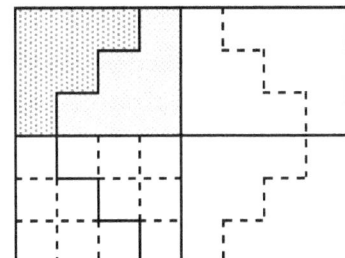

Figure 10a. $4 \times (3 \times 4)$ Figure 10b. $8 \times \frac{3 \times 4}{2}$

Therefore we can conclude that $(2n+1)^2 - 1 = 8 \times \frac{n(n+1)}{2}$. Another way to see that $\frac{n(n+1)}{2}$ is a whole number is because either n or $n + 1$ is an even number.

Example 4. The cube of a number minus the number.
We can represent raising a number n to the third power by forming an actual

cube with n^3 unit cubes. Subtracting the original number from this cube can be done by deleting one column of n unit cubes (figure 11a). One of the incomplete slices can be rearranged to form a rectangular brick (figure 11b). Its dimensions are $(n - 1)$ by n by $(n + 1)$. Therefore $n^3 - n = (n - 1) \cdot n \cdot (n + 1)$. We see that the numbers on the right side are three consecutive numbers. Therefore one of them has to be divisible by 3, and at least one has to be divisible by 2. The product is therefore divisible by 6.

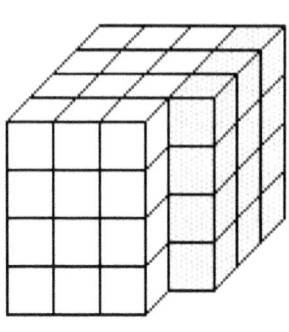

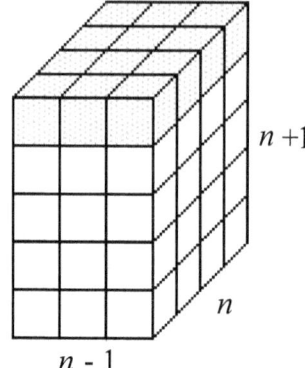

Figure 11a. $n^3 - n$ Figure 11b. $(n - 1) \cdot n \cdot (n + 1)$

Example 5. Sums of odd numbers and sums of cubes.
Students can look at the triangle of odd numbers in table 2. They can compute the sum on each line, and verify that the sum is equal to a cubic number. We can represent the sums in each row in table 2 with the rectangles in figure 12. Students can identify which rectangle corresponds to what row and explain why. They can predict what the next row will be. They can draw the next rectangle. They can relate the numbers in each row with different parts of the corresponding rectangle.
Students can express verbally the suggested relationship. Each rectangle represents, on one hand, the sum of consecutive odd numbers (each odd number is a thin straight or L shaped strip within each rectangle). On the other hand, each rectangle is the cube of a number, because it has n squares of area n^2. Each rectangle has a growing number of strips, 1 the first, 2 the second, 3 the third, and so on. Thus, the total number of strips up to a given rectangle is the corresponding triangular number $(1 + 2 + ... + n)$. For example, the first two rectangles have $1 + 2$ strips, therefore the third rectangle will start with the fourth odd number $7 = 2 \times 4 - 1$, which can also be written as $2 \times 3 + 1$, and it includes three consecutive odd numbers. In general, the n-th rectangle is the sum of n odd numbers, starting with $2(1 + 2 + ... + (n - 1)) + 1$ and finishing with $2(1 + 2 + ... + n) - 1$.

Table 2. Sums of odd numbers

1	1	1^3
3 + 5	8	2^3
7 + 9 + 11	27	3^3
13 + 15 + 17 + 19	64	4^3

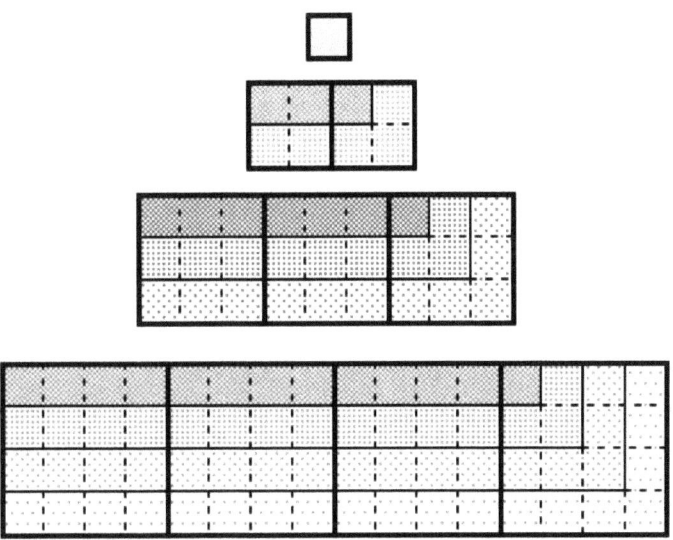

Figure 12. Sums of odd numbers and sums of cubes.

Example 6. Products of three consecutive numbers.

Students may notice that if we have three consecutive numbers (for instance 3, 4, 5), the product of the first by the third ($3 \times 5 = 15$) differs by one from the square of the middle number ($4^2 = 16$). Students can represent this by a rectangle and a square, and overlapping them as in figure 13.

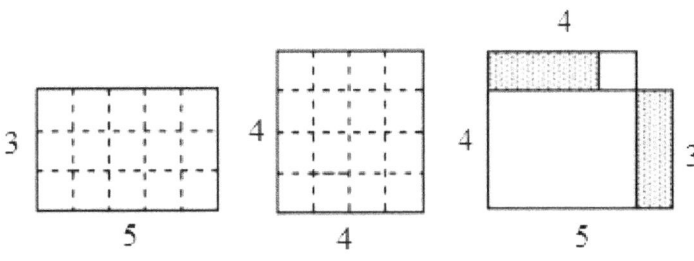

Figure 13. $3 \times 5 + 1 = 4^2$

Students can express the number relations in general terms. The product of two numbers that differ by two, plus one, is equal to the square of their average. Then they can express the relation by using algebraic symbols. If m, $m + 1$, and $m + 2$ denote the three consecutive numbers, then $m(m+2)+1=(m+1)^2$. Figure 14 can help both to see why this general relation is true, and as a guide to give meaning to the intermediate steps when the equation is verified algebraically. For example, as students expand $m(m + 2)$, they can see that the terms $m^2 + 2m$ correspond to a big square and two strips.

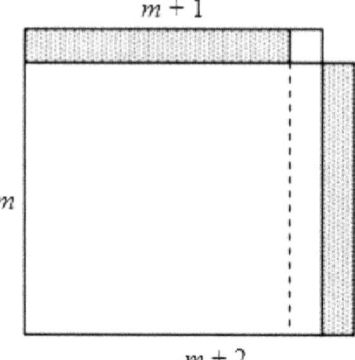

Figure 14. $m(m+2)+1=(m+1)^2$

Example 7. Products of four consecutive numbers.
In a way similar to the example before, if we have four consecutive numbers (for instance 3, 4, 5, 6), and pair the first and the fourth numbers on one hand (3, 6), and the second and the third on the other (4, 5), and form two rectangles whose sides are the pairs of numbers, we will see that their areas differ by two units (figure 15). Students can try other examples.

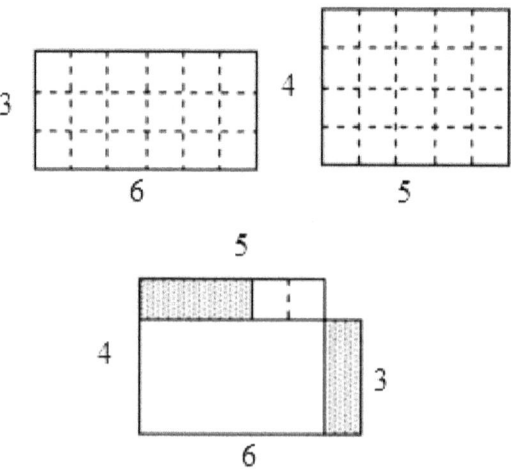

Figure 15. $3 \times 6 + 2 = 4 \times 5$

Students can express the relations among the products of consecutive numbers in general. If we have four consecutive whole numbers, the product of the two middle terms is two more than the product of the first and the fourth. Then they can express the relation in algebraic terms by $n(n+3)+2=(n+1)(n+2)$. They can see why this result is true in general from figure 16. The parts of figure 16 can also provide meaning to the terms in the intermediate steps when the equality is proved algebraically. For instance they can identify $n^2 + 3n$ as one square and three strips.

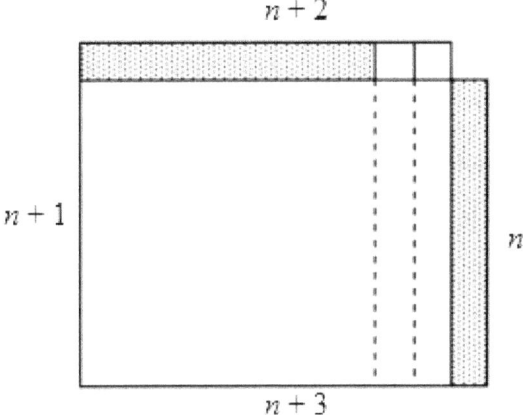

Figure 16. $n(n+3)+2=(n+1)(n+2)$

Example 8. The product of four consecutive numbers.
Let us consider the product of four consecutive numbers. For instance $2 \times 3 \times 4 \times 5 = 120$. The product differs by one from a square number, $121 = 11^2$. Students can relate this to the previous two examples. As we saw in example 7, the product of the first and fourth numbers differ by two from the product of the second and the third. That is, if we let $n(n+3) = m$, then $(n+1)(n+2) = m + 2$. From example 6, $m(m + 2) + 1 = (m + 1)^2$, or using the products $n(n + 3)$ and $(n + 1)(n + 2)$, and rearranging the terms, we have $n(n+1)(n+2)(n+3) = [n(n+3)+1]^2$. So students can see why the product of four consecutive numbers plus one is a perfect square. (Of course, with today's technology, students could simple ask a calculator with algebraic capabilities to factor $(x \cdot (x+1) \cdot (x+2) \cdot (x+3)+1)$ and see that it is indeed equal to $(x^2 + 3x + 1)^2$, but the gained insight of why it is true would be lost.)

Example 9. An inequality
Inequalities can also be represented geometrically. In figure 17 the length of each of the shaded rectangles is a, and their width is b. The two squares

represent a^2 and b^2. Clearly, the two squares cover more area than the two rectangles. Therefore $a^2 + b^2 \geq 2ab$. The side of the remaining white square is $a - b$, therefore $(a - b)^2 = a^2 + b^2 - 2ab$.

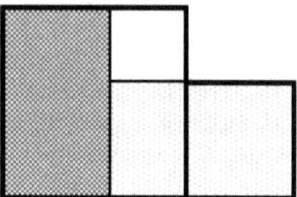

Figure 17. $a^2 + b^2 \geq 2ab$.

Example 10. Sums of squares $1^2 + 2^2 + 3^2 + ... + n^2$

We can represent square numbers by slices that have a square base and one unit height. We can also represent the sum of square numbers by stacking such slices into an echelon building (figure 18). Students can make copies of this building with interlocking cubes (make sure the links do not protrude). If students work in teams, they can easily construct six copies. Each one represents the same sum of square numbers (figure 19).

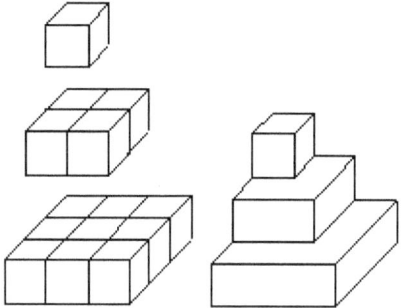

Figure 18. Sum of square numbers $1 + 4 + 9$

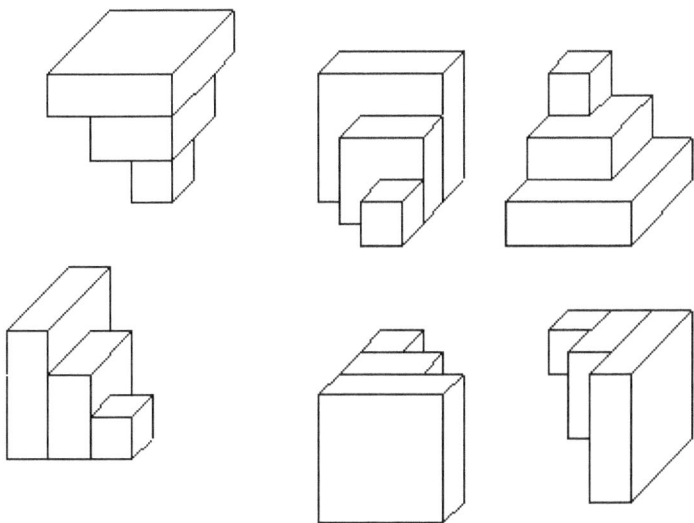

Figure 19. Six sums of square numbers

Students can now arrange the six copies into a block (figure 20). In this case it will have dimensions 3 by 4 by 7. In general, if we add six echelon buildings representing each the sum $1^2 + 2^2 + ... + n^2$, we will form a rectangular block of dimensions n by $(n + 1)$ by $(2n + 1)$. Therefore, the formula to add the first n square numbers is given by

$$1^2 + 2^2 + ... + n^2 = \frac{n(n+1)(2n+1)}{6}.$$

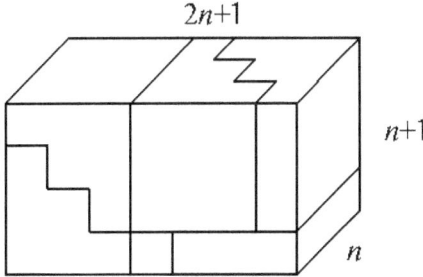

Figure 20. $6(1^2 + 2^2 + ... + n^2) = n (n + 1) (2n + 1)$

Example 11. Sum of triangular numbers
Triangular numbers can also be represented with stairs of cubes. We can use interlinking cubes to represent sums of triangular numbers as shown in figure 21. (Again, make sure that the links do not protrude.)

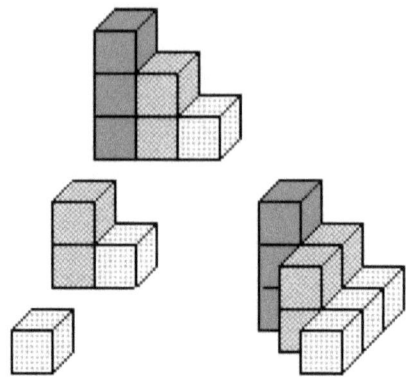

Figure 21. Sum of triangular numbers

Students can make two more copies of the building, one of which is a mirror image of the original (see central part of figure 22). If T_n denotes the n-th triangular number, the sum of the first n triangular numbers is $T_1 + T_2 + \ldots + T_n$. By joining the three buildings to form a stair we can see that three times the sum of triangular numbers is equal to $(n + 2) \cdot T_n$ (left part of figure 22). By arranging the three building to form a wider stair, we can see that the total is given by $n \cdot T_{n+1}$ (right part of figure 22). Because $T_n = \frac{n(n+1)}{2}$, we have in each case that $3(T_1 + T_2 + \ldots + T_n) = \frac{n(n+1)(n+2)}{2}$. Therefore, a formula for the sum of the first n triangular numbers is given by $T_1 + T_2 + \ldots + T_n = \frac{n(n+1)}{2}$.

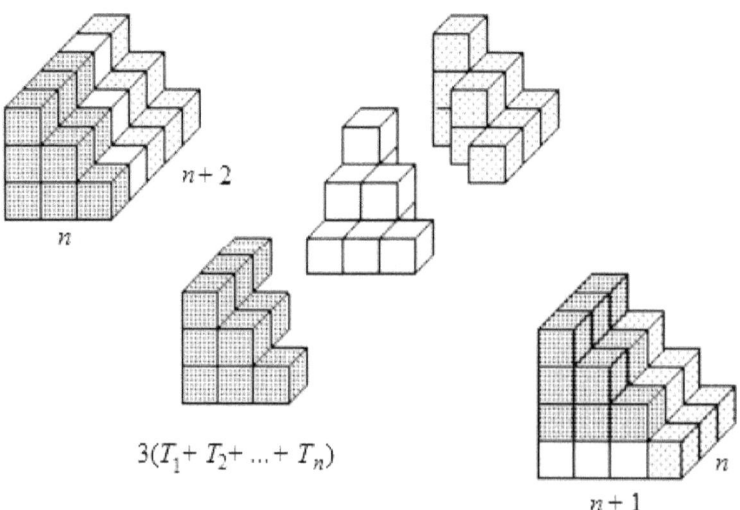

Figure 22. $3(T_1 + T_2 + \ldots + T_n) = (n + 2)T_n = n\,T_{n+1}$

Example 12. Sums of triangular numbers and sums of special products.

By joining one building that represents the sum of triangular numbers with its mirror image, we can describe the sum of triangular numbers as the sum of special products. Notice that the number of cubes in each level of the joint building is given by 1×2, 2×3, 3×4, and in general by $n \times (n + 1)$. We can now use three copies of these double buildings (six sums of triangular numbers), to form a rectangular block. In this case the dimensions of the block will be 3 by 4 by 5, and in general n by $(n + 1)$ by $(n + 2)$. Thus we obtain the same formula for the sum of triangular numbers, or a formula for the sum of special products $1 \times 2 + 2 \times 3 + \ldots + n \times (n + 1) = \dfrac{n(n+1)(n+2)}{3}$.

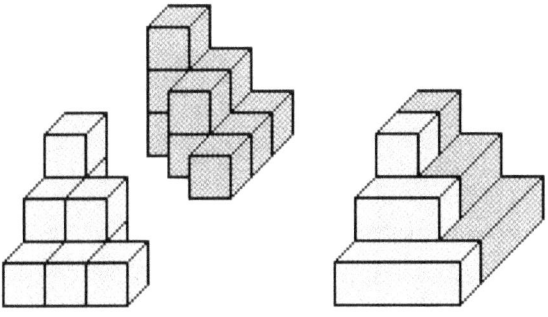

Figure 23. $2(T_1 + T_2 + \ldots + T_n) = 1 \times 2 + 2 \times 3 + \ldots + n \times (n + 1)$

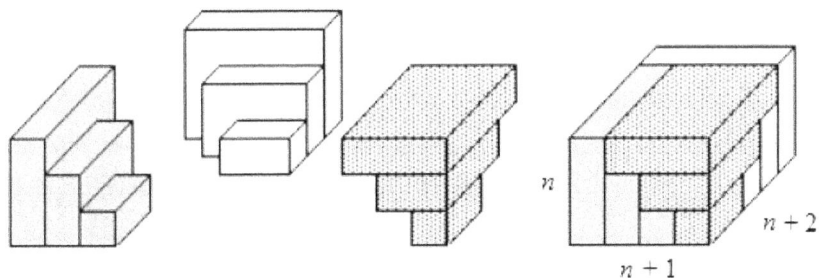

Figure 24. $T_1 + T_2 + \ldots + T_n = \dfrac{n(n+1)(n+2)}{6}$

Conclusion

All too often students are asked to make the transition from arithmetic to algebra too quickly. Students need time, opportunity, and support to make the transition. Concrete representations of numbers and their relations in the form of manipulative materials, puzzles, and visual displays can help students give meaning to the symbolic expressions. In addition, by using concrete displays and particular numbers, students can explore relations that can be generalized. By using and extending the patterns they observe among numbers, they will find it easier to use variables later to generalize the patterns. Polya (1962) points out that, "abstractions are important; use all means to make them more tangible" (p. 102). In the examples shared here, geometric representations of particular numbers were used to make general statements tangible. The representations can be used to support reasoning in general.

The use of variables implies a reasoning that is general. However, many students are not prepared to simultaneously do general mathematical reasoning and use algebraic symbols. They may need to do the reasoning first with particular numbers. By using geometrical representations, students can develop their abilities to reason in general terms before they use symbols for variables. At the same time, by using geometrical representations, they will develop meanings for the algebraic terms that will be used later.

For students, geometric representations can provide fresh meanings and ways to see at a glance why the relations hold. Furthermore, these representations can help students to establish a connection with the algebraic notation. Geometric representations can help students see what each of the terms in an algebraic equation represents. At the same time, the geometric figures can guide students as they do the symbolic manipulation.

Algebraic notation has the power to carry much of the weight of thinking when deriving or proving mathematical results. Its abstractness allows us to forget the meaning of the terms as we manipulate the symbols. However, in the beginning, while students develop their skills with the symbols, it is important that they connect meaning to what the manipulation of symbols represent, so that their manipulation does not become senseless. Students need to avoid what Sowder and Harel (1998) have described as treating "the symbols as though they have a life independent of any meaning or any relationship to the quantities in the situation in which they arose" (p. 5). Geometrical representations can provide guidance and understanding as to why each step or term in a chain of algebraic manipulations is correct. Later, when students have developed skill in manipulating symbols and their algebraic (or symbol) sense, geometric representations can still be helpful. They can be additional sources of discoveries and inspiration.

References

Algebra Working Group. (1998). A framework for constructing a vision of algebra. In *The nature and role of algebra in the K-14 curriculum* (p. 145 - 190). Washington, DC: National Academy Press.

Arcavi, Abraham. (1999). The role of visual representations in the learning of mathematics. In Fernando Hitt and Manuel Santos (Eds.), *Proceedings* of the *21st Annual Meeting of the North American Chapter of the International Group for the Psychology of Mathematics Education*, vol. 1, p. 55-80. Columbus, OH: ERIC Clearinghouse for Science, Mathematics, and Environmental Education.

Bass, Hyman. (1998). Algebra with integrity and reality. In *The nature and role of algebra in the K-14 curriculum* (p. 9 - 15). Washington, DC: National Academy Press.

Botsmanova, M. E. (1972a). The forms of pictorial visual aids in instruction in arithmetic problem solving. In Jeremy Kilpatrick and Izaak Wirzup (Eds.), *Soviet Studies in the psychology of learning and teaching mathematics*, vol. 6, (p. 105-110). Chicago: University of Chicago.

Botsmanova, M. E. (1972b). On the role of graphic analysis in solving arithmetic problems. In Jeremy Kilpatrick & Izaak Wirzup (Eds.), *Soviet Studies in the psychology of learning and teaching mathematics*, vol. 6, (p. 119-123). Chicago: University of Chicago.

Euclid (1956). *The thirteen books of Euclid's Elements* (edited by Thomas L. Heath) New York, NY: Dover.

Fischer, Efraim (1987). *Intuition in science and mathematics: An educational approach*. Dordrecht: D. Reidel Publishing.

Flores, Alfinio. (1992). A geometrical approach to mathematical induction: Proofs that explain. *PRIMUS, 2*, 393-400.

Flores Peñafiel, Alfinio (1999). Las representaciones geométricas como un medio para cerrar la brecha entre la aritmética y el algebra. *Educación Matemática, 11*(3), 69-78.

Flores Peñafiel, Alfinio. (2000a). Demostraciones que explican. *Boletín de FICOM, 4*, 3-4.

Flores Peñafiel, Alfinio. (2000b). Uso de representaciones geométricas para facilitar la transición de la aritmética al álgebra. *Eureka, 15*, 16-21.

Kieran, Carolyn. (1990). Cognitive processes in learning school algebra. In Pearl Nesher and Jeremy Kilpatrick (Eds.), *Mathematics and cognition* (p. 96-112). Cambridge, England: Cambridge University Press.

Kieran, Carolyn, and Chalouh, Louise. (1993). Prealgebra: The transition from arithmetic to algebra. In D. T. Owens (Ed.), *Research ideas for the classroom: Middle grades mathematics* (p. 179 - 198). Reston, VA: National Council of Teachers of Mathematics.

Krutetskii, V. A. (1976). *The psychology of mathematical abilities in schoolchildren*. Chicago: University of Chicago Press. (see p. 326)

Lodholz, Richard D. (1990). The transition from arithmetic to algebra. In Edgar L. Edwards (Ed.), *Algebra for everyone* (p. 24 - 33). Reston, VA: National Council of Teachers of Mathematics.

Mulligan, Catherine H. (1988). Using polynomials to amaze. In Arthur F. Coxford (Ed.), *The ideas of algebra, K - 12* (p. 206 - 211). Reston, VA: National Council of Teachers of Mathematics.

National Council of Teachers of Mathematics. (2000). *Principles and standards for school mathematics*. Reston, VA: The Council.

Nelsen, Roger B. (1993). *Proofs without words*. Washington, DC: Mathematical Association of America.

Polya, George. (1962). *Mathematical discovery*, vol. 2. New York, Wiley.

Presmeg, Norma C. (1999). On visualization and generalization in mathematics. In Fernando Hitt and Manuel Santos (Eds.), *Proceedings of the 21st Annual Meeting of the North American Chapter of the International Group for the Psychology of Mathematics Education*, vol. 1, p. 151-155. Columbus, OH: ERIC Clearinghouse for Science, Mathematics, and Environmental Education.

Sowder, Larry, and Harel, Guershon (1998). Types of students' justifications. *Mathematics Teacher*, *91*, 670-675.

Wagner, Sigrid, and Parker, Sheila. (1993). Advancing algebra. In Patricia S. Wilson (Ed.), *Research ideas for the classroom: High school mathematics* (p. 119 - 139). Reston, VA: National Council of Teachers of Mathematics.

14 PROFOUND UNDERSTANDING OF DIVISION OF FRACTIONS[14]

Introduction

The purpose of this article is to illustrate the type of mathematical knowledge that teachers need so that the teaching of division of fractions is meaningful for the students. We focus on the properties of understanding emphasized by Ma (1999)—connectedness, multiple perspectives, basic ideas, and longitudinal coherence. Few comprehensive discussions of division of fractions are readily available for today's teachers. There are presentations of division of fractions in terms that are meaningful for students in the upper elementary and middle school, using the measurement interpretation of division, or dividing fractions with equal denominators (Armstrong and Bezuk 1995). This is a good start, but it is not enough. Teachers need a complete picture that connects concrete approaches of division with the algorithm of multiply by the reciprocal. They need to understand the role of reciprocals (multiplicative inverses), and the inverse nature of the operations of division and multiplication.

Traditionally, in the United States, division of fractions has been taught often by emphasizing the algorithmic procedure 'invert the second fraction and multiply' with no effort to provide students with an understanding of why it works; in many cases, quests for understanding are discouraged. Many teachers do not realize that the 'invert' refers to the multiplicative inverse, that is, the reciprocal; rather, they see the instruction to invert only as a

[14] Flores, A. (2002). Profound understanding of division of fractions. In B. H. Litwiller (Ed.), *Making sense of fractions, ratios, and proportions, 2002 Yearbook* (p. 237-246). Reston, VA: National Council of Teacher of Mathematics. Copyright National Council of Teachers of Mathematics. Used by permission.

symbol manipulation where the fraction is 'flipped'.

Many of the concepts needed to understand this topic are based on relations among whole numbers that are multiplicative in nature. These concepts need to be developed before the topic is treated. Understanding these relations is important not only for the teachers in the upper elementary and middle school but also for teachers of earlier grades. In addition to Ma's framework and wealth of examples, Kieren's (1992) excellent discussion of rational and fractional numbers was used in preparing this chapter.

Connectedness

Teachers who understand a topic make connections with other mathematical concepts and procedures. They also emphasize more complicated and underlying connections among different mathematical operations and sub domains (Ma 1999). Some of the connections needed in division of fractions are fractions and quotients, fractions and ratios, division as multiplicative comparison, reciprocals (inverse elements), and inverse operations. These connections will help students' learning to be coherent. Instead of learning division of fraction as an isolated topic, students will learn how it fits in a unified body of knowledge. Division of fractions also provides a setting to develop proportional thinking, which is at the core of mathematics in the middle school. Students also have the opportunity to develop their algebraic thinking by dealing with concepts such as inverse operation and reciprocal.

Multiple perspectives

<u>Meanings of division of fractions</u>

To understand division of fractions teachers need to appreciate different meanings such as measurement division, sharing, finding a whole given a part, and missing factors. They need to value various approaches to a solution and different kinds of explanations. They also need to be aware of advantages and disadvantages and the contexts in which each approach tends to be more helpful.

One meaning of division is the measurement interpretation: How many time does one number fit into another? With the help of concrete models of fractions, students can see that $\frac{1}{4}$ fits two times into $\frac{1}{2}$, therefore $\frac{1}{2} \div \frac{1}{4} = 2$. With some guidance, they will also be able to solve problems like $\frac{1}{6} \div \frac{1}{3} = \frac{1}{2}$, and $\frac{1}{2} \div \frac{1}{3} = 1\frac{1}{2} = \frac{3}{2}$ (figure 1). Sewing or cooking are two contexts commonly used in the measurement interpretation of division as in this example provided by a teacher, "You have $1\frac{3}{4}$ cups of flour, and for each batch of cupcakes you need $\frac{1}{2}$ a cup. How many batches can you make?"

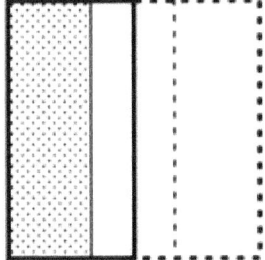

Figure 1. $\frac{1}{3}$ fits one and a half times into $\frac{1}{2}$.

Another meaning is sharing or partitive division. Division as sharing works well when fractions are divided by whole numbers. To solve $\frac{1}{2} \div 3 = \frac{1}{6}$, one half is divided into three equal parts, and the result is $\frac{1}{6}$ of the original whole, for example, when half a cake is shared equally among three children.

One can also interpret division of fractions as a problem of finding a whole given a part. A Chinese teacher explained $1\frac{3}{4} \div \frac{1}{2}$ "means that $\frac{1}{2}$ of a number is $1\frac{3}{4}$. The answer as one can imagine, will be $3\frac{1}{2}$ which is exactly the same as the answer for $1\frac{3}{4} \times 2$. 2 is the reciprocal of $\frac{1}{2}$." (Ma 1999, p. 60)

A problem like $1\frac{3}{4} \div \frac{1}{2}$ can also be interpreted as a missing factor problem: 'What number multiplied by $\frac{1}{2}$ gives $1\frac{3}{4}$?' Although the previous interpretation is similar to this one, they are not exactly the same. The missing factor meaning of division is also used with whole numbers, 2 times what gives 6? However 'finding a whole given a part' does not work for division problems with whole numbers.

For fractions of the same denominator, we can think of them as sets made of pieces of the same size. Students also need to understand the relation between quotients and fractions, that is, $2 \div 3 = \frac{2}{3}$. We can use the meaning of division as a multiplicative comparison between the number of pieces in two sets, that is, as a ratio. When the pieces are the same size, the ratio depends only on the number of pieces involved, not on their size. Thus the problems $2 \div 3$, $\frac{2}{5} \div \frac{3}{5}$, and $\frac{2}{8} \div \frac{3}{8}$ all have the same answer, $\frac{2}{3}$. To divide fractions with same denominators, it is enough to divide the numerators.

Students' methods

Teachers can also learn from students' invented methods, adapt them, and help students see why their methods work. We will present three examples of methods used by students. In the first example, a student (in a class ages

11-12) used an approach that involves inverse proportionality (as well as measurement interpretation). To solve $\frac{1}{2} \div 2$, his thinking was "there is only half a two in one, so there is a quarter in half of one" (Pirie 1988, p. 3).

One Chinese teacher expanded a method used by students: "When working on whole numbers my students learned to solve certain kind of problems in a simpler way by applying the distributive law. This approach applies to operations with fractions too" (Ma, 1999, p. 63). To calculate $1\frac{3}{4} \div \frac{1}{2}$, write $1\frac{3}{4}$ as $1+\frac{3}{4}$, divide each part by $\frac{1}{2}$, and add the two quotients.

A 10-year old girl found a procedure to divide by fractions whose numerator is one less than its denominator using a part-complement representation (Rowland 1997). The procedure can be described symbolically as

$$100 \div \frac{3}{4} = 100 + 100 \div 3$$

$$100 \div \frac{4}{5} = 100 + 100 \div 4.$$

We will use figure 2 to explain this procedure for $100 \div \frac{3}{4}$. The process has two steps. First, we see that $\frac{3}{4}$ goes into 100 units, 100 times (the shaded areas). For each circle, there is a remainder of $\frac{1}{4}$. Then we need to divide 3 pieces of $\frac{1}{4}$ into the remaining 100 pieces of $\frac{1}{4}$. That gives us the additional $100 \div 3$.

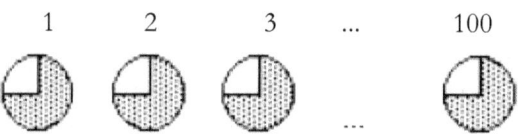

Figure 2. $\quad \frac{3}{4}$ into one hundred

As these examples illustrate, students' thinking can be quite original, and not always easy to understand. Profound understanding of fractions will allow teachers to make sense of these procedures, and help students to make connections to other procedures and concepts.

Ways of justification

Teachers can use concrete representations, empirical evidence and patterns, and properties of numbers and operations to explain the various approaches to division of fractions. Teachers who understand division of fractions can avoid using schemes of justification that are based solely on authority. They will be able to use empirical proofs that go beyond mere perception. They will be able to systematize the examples so that students can see the general nature of the pattern. Furthermore, by using the basic ideas and properties of fractional numbers and operations, teachers will help their students develop analytical schemes of proof.

The teacher can present multiplication problems and division problems that are easy to solve using concrete or pictorial representations, and propitiate the recognition of patterns when students record the results in a systematic way. Students should notice the reciprocals in each row, such as $\frac{1}{3}$ and 3. They can then explore whether the pattern holds for other problems, such as $\frac{1}{2} \div \frac{1}{3}$.

$$2 \div \frac{1}{3} = 6 \qquad\qquad 2 \times 3 = 6$$

$$2 \div \frac{1}{4} = 8 \qquad\qquad 2 \times 4 = 8$$

$$2 \div \frac{1}{5} = 10 \qquad\qquad 2 \times 5 = 10$$

Students can use the particular case of division of fractions when the fractions have the same denominator to develop the procedure for the general case. Given a division problem with fractions such as $\frac{3}{4} \div \frac{2}{5}$ they can find equivalent fractions with the same denominator, and then divide the numerators. Students who know how to multiply fractions will then be able to make the connection to the 'multiply by the reciprocal' procedure,

$$\frac{3}{4} \div \frac{2}{5} = \frac{3\times5}{4\times5} \div \frac{2\times4}{5\times4} = \frac{3\times5}{2\times4} = \frac{3\times5}{4\times2} = \frac{3}{4} \times \frac{5}{2}.$$

Another important special case is when the unit, 1, is divided by a fraction. Students can use measurement interpretation of division to find the answer to problems like $1 \div \frac{2}{5}$. They can see that $\frac{2}{5}$ fits two and a half times into 1. By using also improper fractions, students will be able to notice that the result is the reciprocal of the number they are dividing by, $1 \div \frac{2}{5} = 2\frac{1}{2} = \frac{5}{2}$. Students can then use proportional thinking to find the answer for the general case. If $1 \div \frac{2}{5} = \frac{5}{2}$ then $2 \div \frac{2}{5} = 2\times\frac{5}{2}$, because 2 is twice as big as 1, so the result has to be twice as big. In the same way $\frac{3}{4} \div \frac{2}{5} = \frac{3}{4}\times\frac{5}{2}$, because the result has to be $\frac{3}{4}$ as big.

Basic ideas

Teachers with profound understanding of division of fractions are well aware of the basic concepts and principles of fractions and division, such as identity element for multiplication, reciprocals (multiplicative inverses), and the inverse nature of the operations of division and multiplication. As they teach, they revisit and reinforce these basic ideas. These ideas also guide their ways of justification.

The multiplicative identity element, 1, plays a central role. In the context of rational numbers, the number 1 has a double meaning. The number 1 is the unit of comparison and it is the identity element for multiplication. Multiplying a fraction by 1 does not change the value of the fraction. Students need to see that when multiplying by 1, the identity element is often written as $\frac{2}{2}, \frac{3}{3}, \frac{4}{4}$, etc.

Reciprocals need to be stressed. Students need to realize that $\frac{1}{4}$ is the unit divided into 4 equal parts, and also that the unit is made of four $\frac{1}{4}$. Students also need to see that $\frac{1}{4}$ of 4 is 1. These reciprocal relations hold of course for all fractions. $\frac{5}{2}$ is two and a half units, and also two fifths of $\frac{5}{2}$ is 1. Fractions whose product is 1, such as $\frac{1}{5}$ and 5, need to be connected in the students' minds.

Children go through several stages to develop the idea of fraction in the context of subdividing areas (Piaget, Inhelder, & Szeminska 1960). Teachers need to make sure students have developed a fairly complete understanding of fractions before discussing division of fractions. As Kieren (1992) points out, a full understanding of fractions in this context requires the ability to partition wholes (or units of different kinds), to reconfigure wholes from parts, which is a psychological basis for the concept of inverse, and to subdivide a part and relate the subdivisions to the part but also to the original whole, which is a geometric basis for composite operations.

Rational and fractional numbers are at the same time quotients and ratios (Kieren, 1992). Fractions like $\frac{1}{4}$ and $\frac{2}{8}$ are equal in an absolute sense (extensive quantity) and in a relative sense (proportional equality). Students are faced with results that are simultaneously an amount or the result of a division, for example $\frac{1}{4}$ of a cake as one's share, and a ratio (2 cakes for 8 persons or 1 for 4). Teachers need to understand how the concepts of a fraction $\frac{3}{4}$, a quotient $3 \div 4$, and the ratio 3 to 4 are different and related to each other.

Reversibility of the operations of multiplication and division is central also in the context of fractions. We will discuss the relation between these two operations later.

Longitudinal coherence

The knowledge of teachers with understanding is not limited to what they teach at a certain grade. Rather, they have achieved a fundamental understanding of fractions and their relation to the other areas in the elementary mathematics curriculum; they are ready to take advantage of an opportunity to review crucial concepts previously learned by students, such as equivalent fractions, meanings of division with whole numbers, and the relation between multiplication and division. Teachers also know what topics students are going to learn later that are related to division of fractions, and lay the proper foundation for them. Multiplicative thinking (such as comparison of quantities in terms of ratios) and proportional thinking are crucial for students' success in algebra.

Previous knowledge of division

A thorough understanding of the operations of division and multiplication with whole numbers is basic for understanding division of fractions. Students need to be familiar with several meanings of division, such as sharing and measurement. Whereas with whole numbers usually both interpretations are possible, with fractions, depending on the problem, usually one interpretation is more helpful than the other. Particularly important is that students make the connection between fractions and quotients. Three cakes shared equally by five children will give $\frac{3}{5}$ to each one, $3 \div 5 = \frac{3}{5}$. An important special case is when the unit is divided, $1 \div 5 = \frac{1}{5}$. Sometimes students conceptualize division of whole numbers and fractions as separate and distinct, as one student expressed "we are not talking about fractions, we are talking about dividing" (Toluk 1999, p. 182). Division is also a multiplicative comparison of two quantities, that is, a ratio. The divisions $3 \div 5$, $6 \div 10$, and $9 \div 15$ all have the same result because the ratio in each case is 3 to 5.

Equivalent fractions

Students need a good understanding of equivalent fractions. This includes seeing that the fractions $\frac{1}{2}, \frac{2}{4}, \frac{3}{6}, \frac{4}{8}$, are all equivalent, because they cover the same area in a fraction model, and because the relation of the numerator to the denominator is the same, the numerator is half the denominator. They need to realize that one way to obtain an equivalent fraction is multiplying by 1, written, for example, in the form $\frac{2}{2}$. Students need facility expressing a fraction as a mixed number or as an improper fraction.

Multiplication and division

In the same way that subtraction is the reverse operation of addition, division is the reverse operation of multiplication. If we divide and multiply by the same number, the result is the number that we start with, that is, multiplication "undoes" division, $(8 \div 2) \times 2 = 8$. This reverse relationship is not limited to cases when the quotient is a whole number, $(3 \div 5) \times 5 = 3$. Of course, division also "undoes" multiplication, $(8 \times 4) \div 4 = 8$.

This reverse relationship between multiplication and division holds also for fractions: $\frac{3}{4} \div \frac{1}{4} = 3$, and $3 \times \frac{1}{4} = \frac{3}{4}$, therefore $\left(\frac{3}{4} \div \frac{1}{4}\right) \times \frac{1}{4} = \frac{3}{4}$. So, when a fraction is divided by another fraction and then the result is multiplied by the second fraction, we obtain the original fraction $\left(\frac{3}{4} \div \frac{2}{5}\right) \times \frac{2}{5} = \frac{3}{4}$.

From the last equation we can establish a connection to the "invert and multiply" algorithm. We can multiply both sides by $\frac{5}{2}$ to have $\left(\frac{3}{4} \div \frac{2}{5}\right) \times \frac{2}{5} \times \frac{5}{2} = \frac{3}{4} \times \frac{5}{2}$. The left side is equal to $\left(\frac{3}{4} \div \frac{2}{5}\right) \times 1$ which is equal to $\frac{3}{4} \div \frac{2}{5}$, therefore $\frac{3}{4} \div \frac{2}{5} = \frac{3}{4} \times \frac{5}{2}$. Notice that in this approach $\frac{3}{4} \div \frac{2}{5}$ is treated as a mathematical object, an important step in the transition from arithmetic to algebra.

Students can also see that $\frac{3}{4} \times \frac{5}{2}$ is equal to $\frac{3}{4} \div \frac{2}{5}$ by using the fact that $a \div b = c$ if and only if $c \times b = a$. So, to verify that $\frac{3}{4} \times \frac{5}{2} = \frac{3}{4} \div \frac{2}{5}$, multiply $\frac{3}{4} \times \frac{5}{2}$ by $\frac{2}{5}$ and see that in fact we obtain $\frac{3}{4}$. An important special case for fractions is when reciprocals are multiplied. Because $\frac{a}{b} \times \frac{b}{a} = 1$, therefore $1 \div \frac{a}{b} = \frac{b}{a}$.

The approach of maintaining the value of a quotient underlies the procedure to divide decimals. To compute $0.25\,\overline{)1.75}$, both numbers are multiplied by 100; the answer for $25\,\overline{)175}$ is the same. Likewise, for a problem like $1\frac{3}{4} \div \frac{1}{2}$, when the dividend and the divisor are multiplied by the same number, the result will not change. Thus, $1\frac{3}{4} \div \frac{1}{2} = \left(1\frac{3}{4} \times 2\right) \div \left(\frac{1}{2} \times 2\right) = \left(1\frac{3}{4} \times 2\right) \div 1 = 1\frac{3}{4} \times 2$. This approach is used by some Chinese teachers (Ma, 1999). We can use the same principle with fractions. Because $\frac{1}{2} = 1 \div 2$, we can write $1\frac{3}{4} \div \frac{1}{2}$ $= 1\frac{3}{4} \div (1 \div 2)$. This is equal to $\left(1\frac{3}{4} \div 1\right) \times 2 = 1\frac{3}{4} \times 2$.

Students can use proportional reasoning in division. After solving a sequence of problems like $4 \div 4 = 1$, $8 \div 4 = 2$ and $16 \div 4 = 4$ where the divisor is constant, students will notice that as the dividend increases by a factor of two so does the result. They can realize that the next problem $32 \div 4$ has to be two times bigger (direct proportionality). Likewise, from looking at a sequence of problems like $8 \div 4 = 2$, $8 \div 2 = 4$, $8 \div 1 = 8$, they can see that the dividend is constant and as the divisor becomes twice as small, each result is twice as big as the previous, so that $8 \div \frac{1}{2} = 16$ (inverse proportionality).

Multiplication of fractions
A sound understanding of multiplication of fractions is also necessary. One meaning of multiplication of fractions arises in situations such as $\frac{1}{2}$ of $\frac{1}{4}$. It is important to make the connection of symbol to language, and also know

what are some of the common sources of misunderstanding. Whereas '4 divided in two' is written as $4 \div 2$, '4 divided in half' actually means the same as ' $\frac{1}{2}$ of 4', that is, $\frac{1}{2} \times 4$. The statement $4 \div \frac{1}{2}$ is read as 4 divided by $\frac{1}{2}$. Students also need to understand the area model for multiplication of fractions. If a rectangle has sides that correspond to fractions, its area will be the product of its sides (see figure 3). The area model can also provide a rationale for the procedure 'multiply across'.

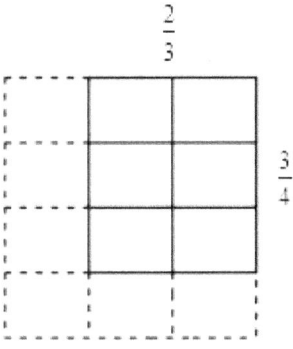

Figure 3. The product $\frac{2}{3} \times \frac{3}{4}$

Compositions of operations

Students need to be able to transform combinations of operations, for example dividing by a quotient of whole numbers, such as $16 \div (8 \div 2) = (16 \div 8) \times 2$. It is important that students learn to deal with a division as a mathematical object, not only as a computation to be done. This is different from the common way to teach about parenthesis: "Compute whatever is in parenthesis first." Other combinations of operations are multiplying by a division, such as $4 \times (6 \div 3) = (4 \times 6) \div 3$, and dividing by a product, such as $16 \div (4 \times 2) = (16 \div 4) \div 2$.

Division of fractions is also equivalent to a combination of operations. Some students discover this on their own. Pirie (1988) reports that in a class with students ages 11-12 working with pictorial representations, one girl saw that dividing by a fraction was the same as multiplying by the denominator and then dividing by the numerator. We can express this in general as $a \div \frac{b}{c} = (a \times c) \div b$. Another way to look at a division by a fraction as a composition of operations is to first divide by the numerator, and then multiply by the denominator, $a \div \frac{b}{c} = (a \div b) \times c$.

Connections to algebra

Several aspects that can be stressed in division of fractions have been identified to help the transition from arithmetic to algebra (Kieran 1990):

- make explicit the procedures to solve arithmetic problems;
- use the equal sign to express a symmetric and transitive relation, rather than just to announce a result;
- consider chains of numbers and operations as mathematical objects and not only as procedures to obtain an answer; and
- pay attention to the method or process and not only to the answer.

Algebraic techniques can help student see analytical proofs from a different perspective. The use of a variable can emphasize that $\frac{1}{2} \div \frac{1}{3}$ is a mathematical object in its own right, $\frac{1}{2} \div \frac{1}{3} = x$. We can also write the relationship as $x \cdot \frac{1}{3} = \frac{1}{2}$, and use algebraic procedures to solve equations to see that $x = \frac{1}{2} \times 3$. Students can also use algebraic notation to write in general terms the rules that they formulated for specific numbers. They can express relations such as $a \div b = \frac{a}{b}, \frac{a}{b} \div \frac{c}{b} = a \div c$, or $1 \div \frac{a}{b} = \frac{b}{a}$.

Students who understand the relationship between division and multiplication, and the role of reciprocals can use different approaches to solve equations. They can think of a problem either in terms of division by a number or multiplication by the reciprocal according to which makes it easier to solve.

Division of fractions is not the only instance where an operation (division) is changed by the inverse operation (multiplication) at the same time that the corresponding number is changed by its inverse (reciprocal). When working with integers, for example, instead of subtracting a number we can add its (additive) inverse, $a - b = a + (-b)$.

Conclusion

As has been documented by researchers, it is not easy for teachers to develop profound understanding of division of fractions and establish the necessary connections on their own. Furthermore, in the United States, division of fractions has been taught for so long by giving a procedure with no explanation why it works, that most of today's teachers and prospective teachers only know that method, so that it is not likely they will learn to teach division of fractions meaningfully from each other. In order to break the cycle of superficial knowledge of procedures, the kind of knowledge described in this chapter needs to become part of the systematic preparation teachers receive either in preservice or inservice courses.

References

Armstrong, Barbara and Nadine Bezuk. Multiplication and division of fractions: the search for meaning. In Judith T. Sowder and Bonnie Schappelle (Eds.) *Providing a foundation for teaching mathematics in the middle grades* (pp. 85-119). Albany, N.Y.: State University of New York Press,

1995.

Kieran, Carolyn. Cognitive processes in learning school algebra. In Pearl Nesher and Jeremy Kilpatrick (Eds.), *Mathematics and cognition* (pp. 96-112). Cambridge, England: Cambridge University Press, 1990.

Kieren, Thomas E. Rational and fractional numbers as mathematical and personal knowledge. In Gaea Leinhardt, Ralph Putnam and Rosemary A. Hattrup (Eds.) *Analysis of arithmetic for mathematics teaching* (pp. 323-371). Hillsdale, N.J.: Erlbaum, 1992.

Ma, Liping. *Knowing and teaching elementary mathematics*. Mahwah, N.J.: Erlbaum, 1999.

Piaget, Jean, Bärbel Inhelder, and Alina Szeminska. *The child's conception of geometry*. New York: Basic Books, 1960.

Pirie, S. E. B. Understanding: instrumental, relational, intuitive, constructed, formalised...? How can we know? *For the Learning of Mathematics*, *8*(3) (1988), pp. 2-6.

Rowland, Tim. Dividing by three-quarters: What Susie saw. *Mathematics Teaching*, *160* (1997), pp. 30-33.

Toluk, Zulbiye. *Children's conceptualization of the quotient subconstruct of rational numbers*. Unpublished doctoral dissertation, Arizona State University, 1999.

15 THE KINEMATIC METHOD IN GEOMETRY[15]

ABSTRACT: Use of arguments of velocity to demonstrate invariance of relationships among segments of geometrical figures.

KEYWORDS: Kinematic method, velocity of endpoints, vectors, geometry, dynamic geometry programs.

INTRODUCTION

The capacity of dynamic geometry programs [5] to trace the locus of points by drawing the position of a point at successive times provides a powerful help to see how the velocities of endpoints of segments are related. With the *Trace Discretely* option on, when the point is dragged quickly, the separation between the points that form its trace will be bigger than when the point is dragged slowly. In Figure 1, C is the midpoint of segment AB. When A is stationary, and AB is stretched and rotated, the magnitude of the velocity of C is half that of B as both move simultaneously. The separation between successive positions of B is twice as big as that of C. The total distance traveled by B will also be twice that of C. In this case the two trajectories are similar, with a dilation factor of 2.

---

[15] Flores, A. (2002). The kinematic method in geometry. *PRIMUS*, *12*, 321-333.
Reprinted by permission of Taylor & Francis (http://www.tandfonline.com).

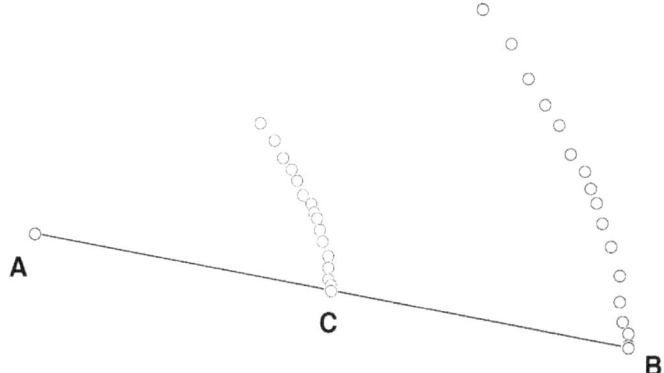

Figure 1. Relations between segments, and velocities of points.

Dynamic geometry programs provide a very convenient setting for students to understand and use the kinematic method, in particular, the theory of velocities. In this paper we will consider the points of geometrical figures as endpoints of changing vectors. We will think of segments that are deformed as changing vectors. As geometrical figures are deformed, some of the relations between its parts will change, while others will remain invariant. In some cases, by proving invariance of the relations between velocities of endpoints we will be able to prove the invariance of the relations of segments of the figure.

Of course it is better if one can deform the figures oneself, rather than just look at a static figure. Using the web component of the dynamic geometric software [6], interactive figures have been posted on the web [3] so that readers have the opportunity to see the figures as they are changed. The number of the interactive figure corresponds to the figure with the same number in this paper.

RELATIONS BETWEEN VECTORS AND VELOCITIES OF ENDPOINTS

We will assume familiarity with the basic results of vector algebra and calculus. Let $\mathbf{r} = \mathbf{r}(t)$ be a vector function, with pole O and endpoint M. If at a time t_0 the vector is equal to $\mathbf{r}_0$, and at t_1 it is equal to $\mathbf{r}_1$, then the vector $\Delta\boldsymbol{\rho} = \mathbf{r}_1 - \mathbf{r}_0$ will be the change of $\mathbf{r}$ in the time interval $\Delta\tau = t_1 - t_0$. The average velocity of the endpoint M for that time interval is given by $\Delta\boldsymbol{\rho} / \Delta\tau$. The instantaneous velocity of the endpoint M of the vector $\mathbf{r}$ at time t_0 is given by

$$\mathbf{v} = \lim_{\Delta t \to 0} \frac{\Delta\boldsymbol{\rho}}{\Delta\tau}.$$

The results about vectors and the velocities of their endpoints presented in this section are very similar to the corresponding results in calculus for functions and their derivatives. In calculus, if a relation between function holds, then the same relation will hold for their derivatives. For example, if $g(x) = 2f(x)$, then $g'(x) = 2f'(x)$. On the other hand, if a relation between derivatives holds, then the same relation holds between the functions, except that we need to add a constant. For example, if $g'(x) = 2f'(x)$, then $g(x) = 2f(x) + k$, where k is a constant.

In the same way, if we know that a relation between two vectors $\mathbf{r}_1$ and $\mathbf{r}_2$ is invariant (as in Figure 1 where AB is twice as big as AC), the same relation between the velocities of their endpoints will hold (the velocity of B is twice as big as the velocity of C). On the other hand, for another pair of vectors DB and DC, if we know that the velocity of the endpoint B is always twice as big as the velocity of endpoint C, (see Figure 2), we will be able to conclude that DB is twice as big as DC, except that we need to add a constant vector $\mathbf{k}$ (in this case a vector parallel to AD).

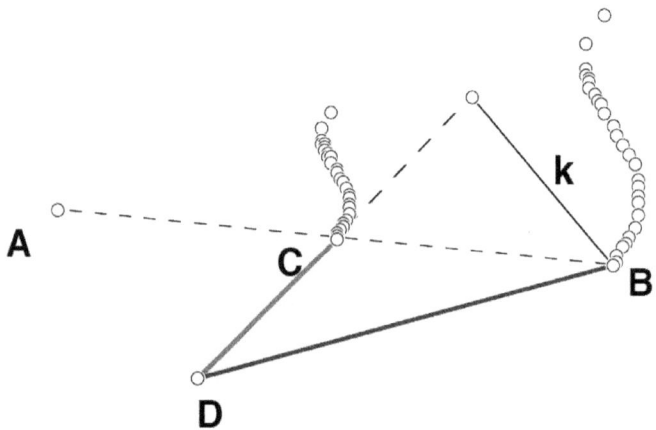

Figure 2. Velocities of endpoints and relation between segments

The next two theorems state precisely how information about the vectors can give us information about the velocities of the endpoints and vice versa. The proofs of the results in this section can be found in [8], or any multivariable calculus book. See [2] for more illustrations of the theorems (as well as more examples of the method described here). Let $\mathbf{r}_1$, $\mathbf{r}_2$ be vectors. The respective velocities of the endpoints will be $\mathbf{v}_1$ and $\mathbf{v}_2$. Let $R_\alpha \mathbf{a}$ be the rotation of vector $\mathbf{a}$ by an angle of α degrees.

Theorem 1. Let m be a constant number and α a constant angle. If the endpoints of vectors $\mathbf{r}_1$, $\mathbf{r}_2$ change in such a way that for all time $\mathbf{r}_1 = mR_\alpha \mathbf{r}_2$, then the velocities of their endpoints are related for all time by the equation $\mathbf{v}_1 = mR_\alpha \mathbf{v}_2$.

Theorem 2. If the velocities $\mathbf{v}_1$ and $\mathbf{v}_2$ of the endpoints of the vectors $\mathbf{r}_1$ and

$\mathbf{r}_2$ are related for all time by the equation $\mathbf{v}_1 = mR_\alpha \mathbf{v}_2$ where m is a constant number and α is a constant angle, then the vectors are related for all time by the equation $\mathbf{r}_1 = mR_\alpha \mathbf{r}_2 + \mathbf{k}$ where $\mathbf{k}$ is a constant vector.

In the following sections we show the value of this view and how these two theorems are used to obtain results in geometry.

EQUILATERAL TRIANGLE ON THREE PARALLEL LINES

Problem. Given three parallel lines, construct an equilateral triangle so that one vertex of the triangle is on each of the three lines.

Solution. Drop part of the condition. Construct an equilateral triangle so that two vertices G and H are on the lower parallel lines; let F be the third vertex of this equilateral triangle (Figure 3).

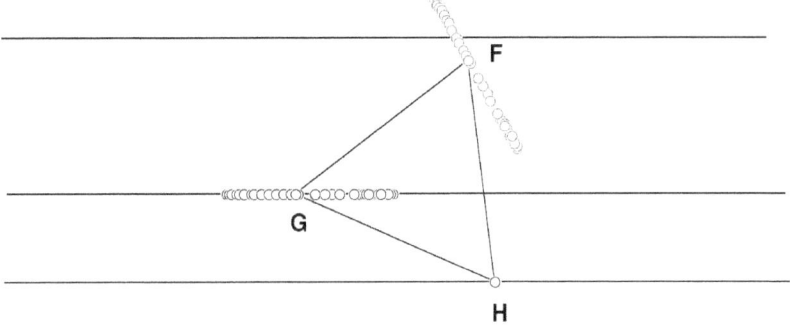

Figure 3. Equilateral triangle on parallel lines.

With H fixed, as point G is dragged along the parallel line and the rest of the figure moving accordingly so that it remains an equilateral triangle, HF always forms a 60° angle with HG, and they are the same length. Therefore the velocity of F will be the same magnitude as that of G but rotated 60° (clockwise in this case). That is, F will move also on a straight line, at 60° with the parallel lines. The desired vertex on the third line will be at the intersection of a line through F making an angle of 60° with the parallel lines and the third parallel.

This example illustrates the use of Theorem 1; knowing a relation about the segments, we know the same relation holds for the velocities of the endpoints. In each of the next examples, both theorems are used. There are three steps involved. First, use the *given conditions* of the problem to ascertain relations between segments of the figures. Then use Theorem 1 to infer the same relations between the velocities of the corresponding endpoints. Then use Theorem 2 to infer the same relation for another pair of segments with the same endpoints. Finally, verify that the constant vector $\mathbf{k}$ of Theorem 2 is zero, by using a *special case* where it is clearly true that $\mathbf{k} = 0$. The general case follows by dragging one or more vertices of the special case figure. In all examples, the proofs stand on their own and do not depend on what we see as we interact with the dynamic figures on the screen. However, the kinesthetic, dynamic use of the figures helps students to have a better and more intuitive understanding of the relations between segments and relations between velocities of endpoints.

EQUILATERAL TRIANGLES ON PARALLELOGRAM

Let ABDC be a parallelogram. ACF and CDJ are equilateral triangles on adjacent sides of the parallelogram (see Figure 4). Then, BJF is an equilateral triangle.

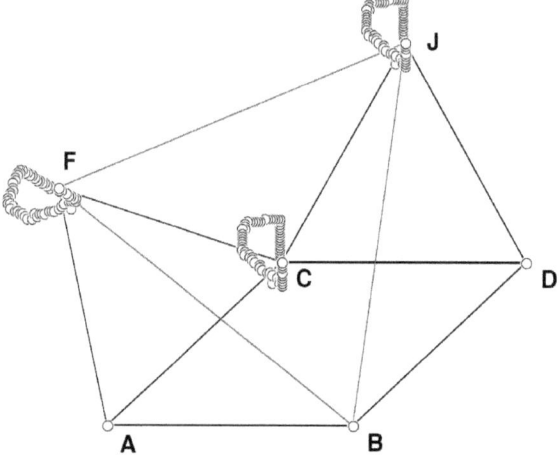

Figure 4. Equilateral triangles on parallelogram.

Proof. Drag point C. Points D and J will have the same velocity as C. Because AC and AF are part of an equilateral triangle, the velocity of F is the same as the velocity of C, rotated 60° (Theorem 1). Therefore the velocity of F is the same as the velocity of J, rotated 60°. Therefore BF is equal to BJ rotated 60°, plus a constant vector **k** (Theorem 2). To see that **k** = **0**, let C coincide with A. HF will be equal to HA, and HD will be equal to HB, and use the given condition that HA and HB are legs of an equilateral triangle.

ASYMMETRIC PROPELLER

Three congruent equilateral triangles share a vertex (see Figure 5). Then the midpoints of the segments connecting the vertices of the triangles form also an equilateral triangle LMN [4]. Proof. Drag D (it will move around the circle so that the condition is satisfied). Because C and D are vertices of an equilateral triangle, the velocity of C will be the same magnitude as the velocity of D and will be rotated -60°. Because M is the midpoint of segment

FD, with F fixed, the velocity of M is 1/2 the velocity of D. In the same way, the velocity of L is 1/2 the velocity of C. Therefore the velocity of M is equal to the velocity of L rotated -60°. Therefore NM = rot 60 NL + **k**. To see that **k** = 0, start with the three congruent equilateral triangles equally spaced in the circle. In that particular case, because of the 60° symmetry of the figure, it is clear that NML is indeed an equilateral triangle.

The result is true even if the original equilateral triangles are not the same size (see Figure 6). Readers can easily adapt the previous argument.

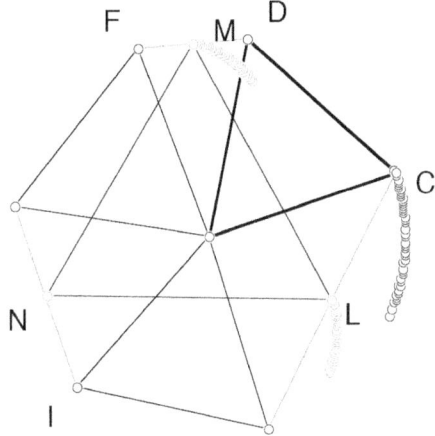

Figure 5. Asymmetric propeller.

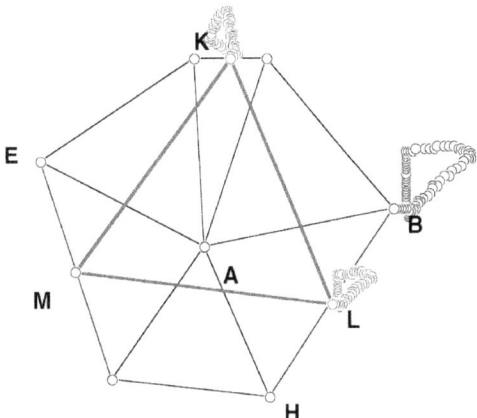

Figure 6. Propellers not the same size.

Furthermore, the result can be generalized for similar triangles of any shape. Let three triangles that are similar to each other (bold triangles in Figure 7) share the appropriate vertices at B as illustrated. Then the resulting triangle KLM formed by the midpoints of segments joining corresponding vertices of the similar triangles will be similar to the original triangles.

Proof. Drag D. Velocity of L is 1/2 velocity of D. Velocity of K is 1/2 velocity of G. Because triangle BGD remains similar, velocity of G is equal to BG/BD rot(DBG) velocity of D. Therefore velocity of K = BG/BD rot (DBG) velocity of L. Therefore MK = BG/BD rot (DBG) ML + **k**. To see that **k** = 0 choose original similar triangles congruent with corresponding sides parallel. Therefore MKL is similar to original triangle

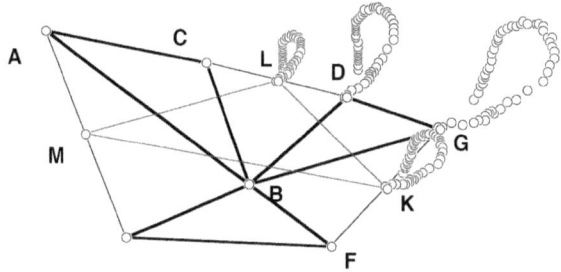

Figure 7. Propeller similar triangles

VECTORS IN A COORDINATE SYSTEM

For the following two examples, O will be the origin located at an arbitrary place on the plane. We identify the point P with the vector OP. We will denote the sum of two vectors OA + OB simply as A + B. We will also use the principle that if the velocities of the endpoints are in opposite directions, and the magnitude of one is n times bigger than the other, then the point that divides the segment into two segments with a ratio 1 to n is fixed.

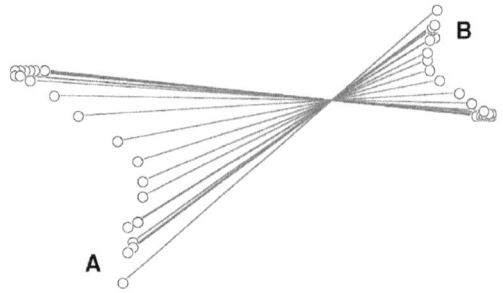

Figure 8. Fixed point.

Centroid of a triangle. In a triangle with vertices A, B, C the point (A + B + C) / 3 is fixed (see Figure 9). That is, it does not depend on the location of the origin, but only on the position of the three vertices of the triangle.

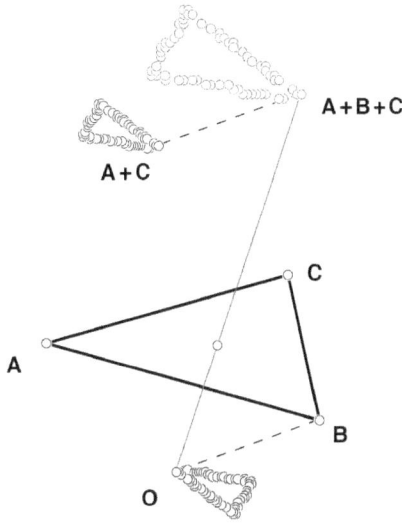

Figure 9.   (A + B + C) / 3

Proof. Drag O. The velocity of A + C is of the same magnitude as the velocity of O, but with opposite direction, because A + C is the reflection of O around the midpoint of AC. The velocity of A+B+C is equal to the velocity of A+C plus the opposite of the velocity of O. Therefore, the magnitude of the velocity of A+B+C is twice the magnitude of the velocity of O, but in opposite direction. Therefore the point at 1/3 the distance between O and A+B+C is fixed. (By the way, this is the centroid of the triangle.)

A center for a quadrilateral. In a quadrilateral ABCD, the point (A+B+C+D)/4 does not depend on the position of O (Figure 10).
Proof. Drag O. The magnitude of the velocity of A+B+C+D is the sum of the velocities of A + D, B+ C, and the opposite of the velocity of O. Thus the velocity of A+B+C+D has three times the magnitude of the velocity of O, and has opposite direction. Therefore the point (A+B+C+D)/4 remains fixed. (This point is also the point of intersection of the medians of the quadrilateral.)

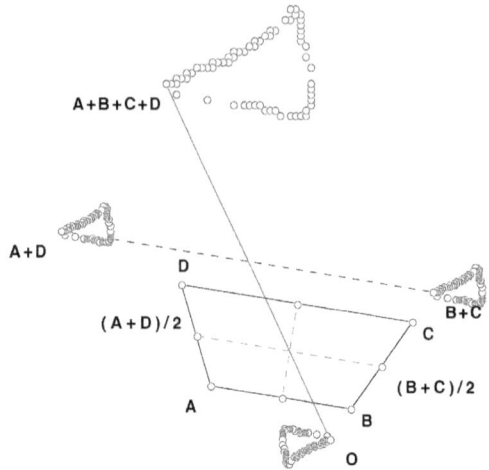

Figure 10. (A+B+C+D) / 4

CENTROIDS OF THE FOUR TRIANGLES OF A QUADRILATERAL

In the following example we will consider a quadrilateral and the centroids of the four triangles associated with it. We will prove first a useful lemma.

Lemma. Let A move with velocity **v**, then the velocity of the point of intersection of the segments connecting the midpoints of opposite sides is 1/4 **v**. (This point of intersection is also the midpoint of each of the two segments).

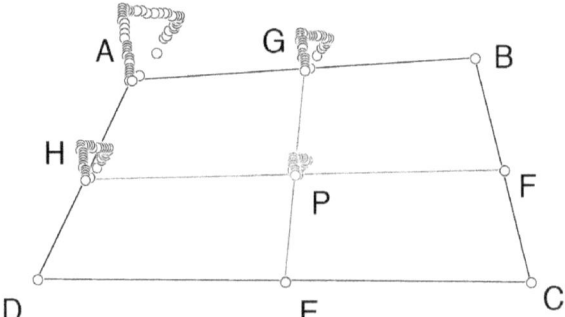

Figure 11. Joining the midpoints of opposite sides.

Proof. Drag A. The velocity of G has the same direction as the velocity of A, but half the magnitude. The velocity of the midpoint of segment GE has the same direction as the velocity of G, but one half the magnitude, so that its velocity is 1/4 that of A.

Remember that two figures are homothetic if they are similar and have a center of projection. We are now ready for the main result in this section.

Let ABCD be an arbitrary quadrilateral. Let GLIM be the quadrilateral formed by the centroids of the four triangles BCD, ABC, ABD, and ACD (Figure 12). The two quadrilaterals are homothetic (ratio -1/3) and the center of homothecy is the intersection of the lines connecting midpoints of opposite sides of the original quadrilateral.

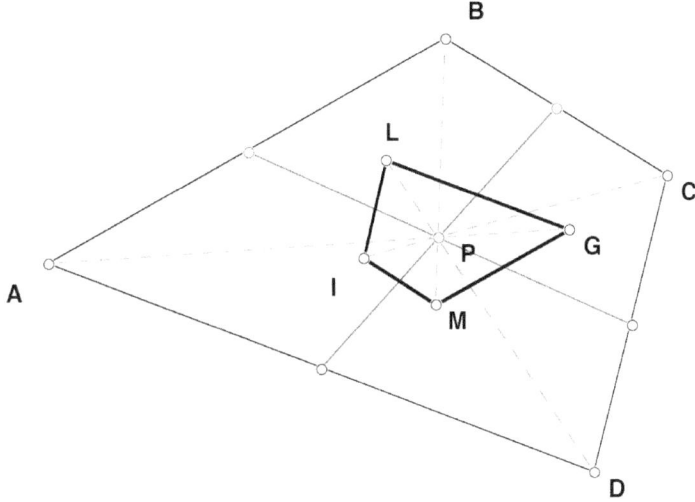

Figure 12. Homothetic quadrilaterals.

Proof. Drag B. Let P the point that divides the segment MB in the ratio 1 to 3. The velocity of P is 1/4 the velocity of B. The velocity of the intersection of the lines joining the midpoints of opposite sides of the quadrilateral is also 1/4 the velocity of B. To show that they are in fact the same point start with a quadrilateral where this is clearly the case (for example a parallelogram). The general case can be obtained by dragging one or more vertices.

PARHEXAGON

On an arbitrary hexagon construct six centroids of the triangles formed by consecutive vertices such as ABC (Figure 13). The six centroids form a hexagon with three pairs of opposite sides that are congruent and parallel sides. This is called a parhexagon [7].

Proof. Drag F. The velocities of centroids G, L, and K, are 1/3 the velocity of F, in the same direction. Therefore segments GL moves parallel to itself. (LK also moves parallel to itself.) Because the velocity of G is the same as the velocity of K then segment HG = RK + **k** (Theorem 2). To see that the segments are parallel to their opposite sides, and that they are congruent, that is, **k** = 0, start with a hexagon where this is clearly true (for example, a regular hexagon).

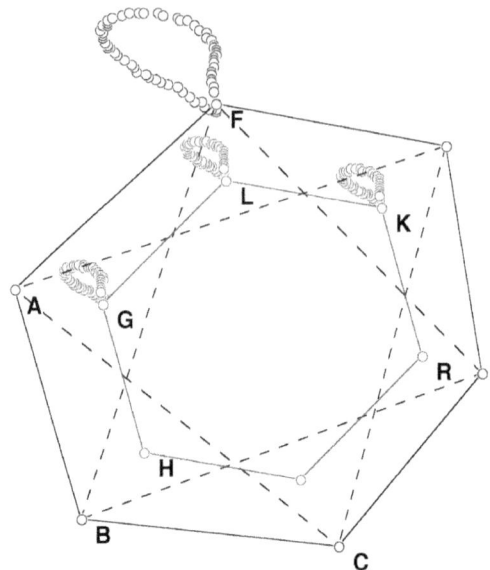

Figure 13. Parhexagon.

CONCLUSION

The results presented here can, of course, be proved using other strategies and tools. The reader may want to contrast the proofs obtained with the kinematic method with other elegant solutions [1]. As teachers of mathematics, we want to encourage students to be able find alternative ways to solve problems, and establish connections between fields such as calculus and geometry. Using the method presented here, students can use velocity to demonstrate invariance of relationships, rather than arguments of static position, length, or measurement.

Students found activities using the kinematic method "intriguing and challenging." It provided the opportunity to discover mathematics in an unfamiliar context. The dynamic geometry program played an important role to make the kinematic method easier to grasp. As one student said, "the trace at the end of the vectors helped compare and contrast two different vectors by their velocities." Another wrote "the greatest asset of both these things, the Kinematic Method and Geometer's Sketchpad is the ability to use them together to get a complete visual picture of the velocities." The interactive figures were also important. As one student phrased it: "By interacting with the figures the learning was made active in that I could make my own conjectures before I read what the answer was.... It is amazing how helpful the sketchpad can be when exploring new concepts, like velocity, and in proving the difference in velocity for different given shapes, points, lines, and angles." Several students made reference to the fact that "the study of vectors and motion has always been in a static environment" and that a visual representation like this could help students in beginning physics courses.

Of course, the method can be applied to other examples. Students may want to try to provide their own proofs for the cases illustrated in the interactive figures 6a, 7a, and 7b [3].

The kinematic method is not new and students can use it without a dynamic geometry program. However, the possibility of *varying the data*, of experimenting, and interacting can help student to better comprehend the relations between vectors and their velocities that are used in the proofs.

REFERENCES

1. DeTemple, D. and S. Harold. 1996. A round-up of square problems. *Mathematics Magazine, 69*, 15-27.
2. Flores, A. 1998. The kinematic method and the Geometer's Sketchpad in geometrical problems, *International Journal of Computers for Mathematical Learning*, 3: 1-12.
3. Flores, A. 2000. Interactive figures for *The kinematic method in geometry,* [On line] available:
   http://www.public.asu.edu/~aaafp/mkinematicmethod.html
4. Gardner, M. 1999. The asymmetric propeller, *The College Mathematics Journal,* 30: 18-22.
5. Jackiw, N. 1995. *The Geometer's Sketchpad* [computer program]. Berkeley, CA: Key Curriculum Press.
6. Jackiw, N. 1998. JavaSketchpad [On line]. Available
   http://www.keypress.com/sketchpad/java_gsp/index.html
7. Kasner, E and Newman, J. R. 1940. *Mathematics and the imagination.* NY: Simon and Schuster.
8. Lyúbich, Yu. I. and L. A. Shor. 1978. *Método cinemático en problemas geométricos.* Moscow: Mir.

BIOGRAPHICAL SKETCH

Alfinio Flores has degrees in mathematics from UNAM in Mexico, and a Ph. D. in Mathematics Education from The Ohio State University. Currently he is Professor of Mathematics Education at Arizona State University. One of his interests is to explore ways in which technology can help make mathematics more interesting and engaging for a broader segment of the population.

16 IF PI WERE EQUAL TO 3...[16]

If pi were equal to 3, the circumference of a circle would be equal to the perimeter of the inscribed regular hexagon. Stating this more precisely, if have a regular hexagon (with perimeter p) inscribed in a circle of radius r (see figure 1), and use the ratio of the perimeter of the hexagon to the diameter, $p / 2r$, to approximate the ratio of the circumference to the diameter $c / 2r$, the approximated value would be 3.

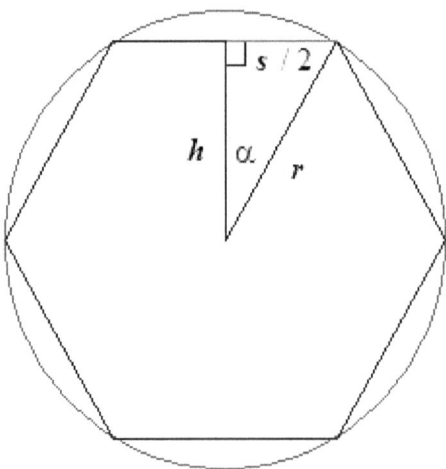

Figure 1. Hexagon inscribed in a circle.

---

[16] Flores, A. (2002). If pi were equal to 3.... *Ohio Journal of School Mathematics*, *46*, 41-44. Reprinted by permission.

Pi is not only the ratio of the circumference to the diameter, $\pi = c / 2r$, it is also the ratio of the area of the circle to the radius squared, $\pi = A / r^2$. So in principle, if we have a regular polygon inscribed in a circle, we could also approximate π by the ratio of the area of the polygon to the radius squared. Students can explore and contrast these two ways to approximate π using circles with inscribed regular polygons, one using the perimeter and the other using the area. In each successive approximation, the number of sides of the polygon is doubled. Students can visually gauge how good is the fit of inscribed polygons that correspond to approximation values of 3, 3.1, 3.14, and so on. An interesting side discovery is that by comparing the two sequences of approximations students will have the opportunity to see how sine and cosine of an angle are related to sine of twice the angle.

For the regular hexagon, the perimeter is simply $6r$ because it is made out of six equilateral triangles. Thus $6r/2r = 3$. To compute the area of the hexagon, we can compute the area of each of these six triangles and then multiply by six. The area of each triangle is $h \times s/2$, where h is the perpendicular from the center to the side, and s is the side (Figure 1). The angle α is obtained by dividing $360 \div 6$ (the number of sides) and then dividing by 2, so $\alpha = 30°$. Therefore $h = r \cos 30°$. The area of the hexagon is thus given by $A = 6 \times s/2 \times h = 6 \times r^2 \times \sin 30° \cos 30°$. Using a calculator that displays 10 digits for values of functions and results of operations, but carries on the operations internally with 14 digits, we obtain the ratio $A / r^2 = 2.598076211$. We also observe that $s/2 = r \sin 30°$. Another way to compute the perimeter is $6 \times 2 \times r \sin 30°$. This method has the advantage that it can be used with other polygons.

For regular dodecagons (12 sides; see figure 2), the values of the perimeter and the area are calculated in the same way as above, but this time the polygon would be formed by 12 isosceles triangles, with common vertex at the center, and $\alpha = 15°$. Therefore $s/2 = r \sin 15°$, and $p = 12 \times 2 \times r \times \sin 15°$; from here we obtain the value $p / 2r = 3.105828541$. For the dodecagon, $h = r \cos 15°$, and $A = 12 \times s/2 \times h = 12 \times r^2 \times \sin 15° \cos 15°$. Using the calculator we obtain $A / r^2 = 3$.

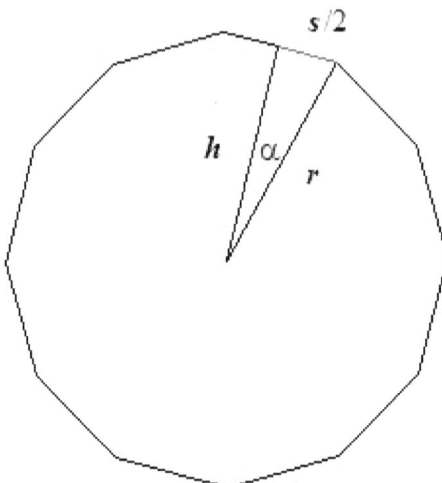

Figure 2. Polygon with 12 sides

A nice surprise for students is to find that sin 15° × cos 15° = 0.25, a rational number. Some students may recall that sin 30° is precisely 0.5. Is this just a coincidence? Students should also notice that the ratio A / r^2 for a polygon of 12 sides is 3, exactly the same value as the ratio p / $2r$ for the hexagon. Another coincidence?

For polygons with 24 sides, we can compute this ratio p / $2r$ the same way as before, using α = 7.5°, to obtain an approximated value of 3.132628613. We can also compute the ratio A / $r2$, which turns out to be 3.105828541. Notice that this is precisely the ratio p / $2r$ for a polygon with 12 sides. This would convince most students that we are not dealing with a simple coincidence, and that some closer look at the issue is warranted. Remember that for the 12 sided polygon, the ratio p / $2r$ is given by 12 ×2 × r ×sin 15°/$2r$ = 12 sin 15°. For the 24 sided polygon, the ratio A / $r2$ is given by 24 × r^2 × sin 7.5° cos 7.5° / $r2$ = 24 sin 7.5° cos 7.5°. Because the two ratios are equal, students can infer that sin 15° = 2 sin 7.5° cos 7.5°. Of course, students who are already acquainted with the formula sin 2α = 2 sin α cos α will recognize this as a special case. Students can continue the process for polygons with 48 and 96 sides. Table 1 displays the values and approximations for the different polygons.

Table 1. Approximated values to π

Number of sides	α	Perimeter / $2r$	Area / r^2
6	30°	3	2.598076211
12	15°	3.105828541	3
24	7.5°	3.132628613	3.105828541
48	3.75°	3.139350203	3.132628613
96	1.875°	3.1411452472	3.139350203
N	$\beta = 180°$ $/ n$	$n \sin \beta$	$n \sin \beta \cos \beta$
$2n$	$\beta/2 \quad =$ $180° / 2n$	$n \sin (\beta/2)$	$2n \sin (\beta/2)$ $\cos (\beta/2)$

As we noticed, the value of the ratio A / r^2, corresponding to a polygon of $2n$ sides, is equal to the ratio $p / 2r$ for polygons with only n sides. Therefore, to approximate π it is better to use the perimeter of a regular polygon than to use its area. Using the area, we would need twice as many sides for the same degree of approximation.

Students will notice that the ratio A / r^2 gets closer to the ratio $p / 2r$ as the number of sides of the regular polygon increases. We can make explicit to what extent the ratio for the area falls short. Remember that a way to compute the area of a regular polygon is $\frac{1}{2} p \times h$, where p is the perimeter and h is the apothem (the perpendicular from the center to one side; that is, $h = r \cos \alpha$). So, $\frac{A}{r^2} = \frac{\frac{1}{2}ph}{r^2} = \frac{\frac{1}{2}p \cos \alpha}{r^2} = \frac{p}{2r}\cos \alpha$. Thus the ratio $\frac{A}{r^2}$ will fall short by a factor $\cos \alpha$ compared with $\frac{p}{2r}$. Because $\cos \alpha$ tends to 1 as α tends to 0, students can see that for a circle the two ratios $c / 2r$ and A / r^2 are equal. Thus students can see why the constant π that appears in the formula for the circumference of the circle, $c = 2 \pi r$, also appears in the formula for the area of the circle, $A = \pi r^2$.

Exercise. Compute the perimeter p of an equilateral triangle inscribed in a circle of radius r. Compute the ratio $p/2r$. Verify that this is the same value as the ratio A/r^2 for the hexagon.

Conclusion
It is important that students realize there are consequences of approximating π to a certain number of decimal places. An approximation of 3 is obviously very coarse, because the perimeter of the inscribed hexagon is clearly different from the circumference. In terms of area, an approximation of 3.1 would put us in the realm of polygons with 24 sides, and working with 3.14

would mean we are approximating the area of the circle with the area of a polygon with about 96 sides. It is very hard to distinguish a figure of 96 sides from a circle, so students can see that for most practical purposes the approximation 3.14 is quite sufficient. However, they need to realize that 3.14 is just an approximation to π, not the real value, in the same way a 96 sided polygon is not a real circle. Students may also find interesting to learn about the different methods used and the approximate values of π attained in different cultures and ages (Von Baravalle, 1989; Beckman, 1970), or about the powerful methods used nowadays (Arndt and Haenel, 2000).

The formula $\sin 2\alpha = 2 \sin \alpha \cos \alpha$ is often presented to students out of the blue; they are left with no idea of how people could think of the formula in the first place. One advantage of the approach described here, it that it provides a context where particular cases of this formula appear. Students can verify it empirically, and familiarize themselves with it. Later, a general proof of the formula can be given.

References

Arndt, Jörg and Christoph Haenel. π *unleashed*. Berlin: Springer Verlag, 2000.

Beckman, Petr. *A history of π*. NY: St. Martin, 1976.

Von Baravalle, Hermann. "The number π." In *Historical topics for the mathematics classroom*. 2nd edition, p. 148-153. Reston, Va.: National Council of Teachers of Mathematics, 1989.

17 HOW DO CHILDREN KNOW WHAT THEY LEARN IN MATHEMATICS IS TRUE?[17]

> Because that's the way all my teachers taught, and those are just
> math rules to get the correct answer for what you are doing. I also
> have faith in my teachers that they are teaching me the right thing.
>
> A 14 year old girl (as reported by SA)

Acknowledgement

This article is based on interviews conducted by
Ismar Adan, Sean Arteaga, Tacy Anderson, Ernestine Aragón, Cheryl Butler,
Jennifer DuPont, Michelle Fenno, Tracy Friddle, María Elena García, Audra
Guest, Almasol Herrera, Cynthia Hesser, Abbie Knutsen, Denis Lawton, Lisa
Meadows, Brian Merrill, Brenda Miller, Lisa Piaskowy, Amy Porter, Keven
Powers, Vicky Rodríguez, Keirsten Russell, Cathy Santo, Julie Starcher,
Andrea Theiler, Cheryl Travelstead, and Keri Vliek.

Introduction

One of the distinctive marks of mathematical thinking is that we do not need
to rely on memory or authority to know whether an assertion is true or not.
In mathematics, we prove our theorems. For students the road to formal
proofs is a long one. The kind and amount of convincing evidence, or proof,
that is appropriate for a child changes as the child's mathematical thinking
evolves. It is important that very early on children develop the attitude and
disposition that they do not need to rely on authority or rote memory to
know why what they learned in mathematics is true.

---

[17] Flores, A. (2002). How do children know what they learn in mathematics is true?
Teaching Children Mathematics, *8*, 269-274. Copyright National Council of Teachers
of Mathematics. Used by permission.

However, as the quotation at the beginning illustrates, students trust their teachers and do not question whether what they learn in mathematics is true. Many do not develop ways to explain why what they learn is true. To help students to think as mathematicians, teachers need to recognize and foster children's ability to verify for themselves if an answer is correct, whether a procedure will produce right answers, or if a formula makes sense. This process should be built on the ways that children use naturally. Therefore, it is important to find out what are the ways that children use to determine whether something they learned in mathematics is true or not.

In this paper we describe some of the ways in which children show why something they know is true. First, we will share a few samples of students' answers, to convey the variety of responses among children, and even between answers from the same child. Then we will group answers by categories of justification that correspond in general terms to the schemas of justification identified by Sowder and Harel (1998). We will briefly describe what teachers can do to help students progress in each category.

The interviews

Students in the mathematics methods course at San Goloteo State University, interview children to find out what they have learned in math, and how they know it is true. First, they ask each child four things that he or she has learned in mathematics. These should be specific facts, such as "two and one is three," not just generic answers such as "addition." They request examples if the answer did not include one. Interviewers write the answers down, using the student's words. Then, in the second part, they ask how the child knows it is true. For example if the child says "eight times zero is zero", they ask how he or she knows the answer is correct. With each explanation, the initials of the interviewer who recorded it are given. The names of the children have been changed.

<u>A sample of children's answers and explanations</u>
Molisto, age 6, showed that $5 + 5 = 10$ by counting on fingers: "I got five"—holding up both hands and counting from one to 10, left to right on his fingers, following with his eyes only. He also said that one hundred and one hundred equals two hundred, explaining: "I made it up." He also stated that one hundred plus a million is one hundred million, explaining also "I made it up." (KP)
Mayra (age 8) explained how she knew that $9 + 5 = 14$, "Vi que faltó uno al nueve para hacer el diez, luego con cuatro son catorce" [I saw that one was missing from nine to make ten, then with four that is fourteen]. Her explanation for $2 + 1 = 3$ was, "Ya lo sé de memoria" [I already know it by heart]. For $5 + 9 = 14$ "Porque ya lo había hecho muchas veces." [Because I had already done it many times]. To show $5 + 6 = 11$, she counted six on

her left hand by holding it open and then folding down five fingers one at a time and then opening and folding down one, then opening hand and counting 7, 8, 9, 10, 11 by folding down each finger, left to right. (KP)

Sean (kindergarten) said to AK as they were looking at the shape of a triangle (fig. 1a). "Triangles all have three sides." AK drew a picture of a different triangle on the chalkboard (fig. 1b). Sean told her that it wasn't a triangle because it didn't look like the first. Sean told AK that squares have four sides. He showed that he knew the sides of squares and triangles by counting out loud the number of sides. Sean added "If I had two squares that would equal to eight sides. If I had two triangles then I would have six sides."

Figure 1a   Figure 1b

Jordan (age 7) stated facts like 50 + 54 = 104, and 10 + 4 = 14. When asked how he knew they were true he answered "For the small addition problems you can use your fingers to count or you can count rocks or other things." For the problem 50 + 54 = 104 he said that 50 + 50 = 100, so then you add four more and you have 104. (AP)

Heather (6 years, first grade) knows the following facts, "10 + 10 + 10 = 30, 10 + 10 + 10 + 10 = 40, 10 + 10 + 10 + 10 + 10 = 50." VR asked her why this was true. She sat there quietly and did not say anything. Then VR asked how she learned those things. She said, "My mom said so." Heather also stated, "Six plus six equals twelve, and five plus five equals ten." She said she knew that because "you can use fingers to help count." (VR)

Children verify empirically, using concrete objects.

At first, children approach mathematics is a very concrete way. They use objects, including their fingers, as tools to help them figure out what the answer is. Bradley, age 6, showed that 2 + 3 = 5, "I put two fingers up"—left hand—"plus 3"—right hand—"and I count 'em"—touched fingers to lips as he counted each one. He stated that 3 + 4 = 7. When he was asked 'How do you know?' he said, "'cuz my fingers, watch"—started with 4 on right hand 1, 2, 3, 4 touching his fingers to his cheek; then 5, 6, 7 on left hand, same touching his fingers to his cheek. For 4 + 3 = 7 the explanation was "I put fingers up and counted them easy"—but without touching to cheek, just looked and spoke the answer with total confidence. (KP)

Children use also visual evidence to know that something is true. Sabrina (kindergarten) told AK that the yellow shape (fig. 2a) was the same as two red together (fig. 2b). She had the pattern of a playground where the outer edge was yellow and she filled it with the red shapes (AK) . Another first grade child also used visual evidence to prove "Four little squares can make a big square" (BMe).

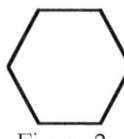
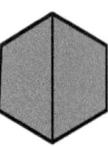

Figure 2a   Figure 2b

Children use counting strategies

Children develop counting strategies to solve addition and subtraction problems. These strategies gradually rely less on concrete objects. Amber was in a combined 2nd-3rd classroom, working with different combinations of animals that would add to five. KV observed that Amber did not use her fingers, but instead she would start with one number and count with a slight nod and the following numbers spoken under her breath. KV asked Amber how she knew that 2 chickens and 3 pigs would equal five animals. Her explanation was, "I said 2 in my head, and then added 3-4-5 on my fingers." Vincent, 3rd grade, uses 'doubles' facts to derive related results. Since he knows 6 + 6 = 12, then 6 + 7 = 13, because 7 is one more than 6. Also, 6 + 5 = 11 because 5 is one less than 6, which makes the answer, 11, one less than 12 (BMi). Courtney, 3rd grade, knows that 7 + 3 = 10, so 7 + 4 = 11 because 4 is one more than 3, which increases the answer by one. In the problem 11 + 3 she puts the big number in her mind and counts on the smaller number. When subtracting, Courtney counts backwards; as in 6 - 3 = 3, she puts the 6 in her mind and counts backwards by one, three times. She knows this is true because if 3 + 3 = 6, then 6 - 3 = 3 (BMi).

Some of the children, as they start working with bigger numbers, develop more efficient counting strategies. For example, a 4th grade student said "If a man goes into a store to buy a $140 computer and gives the cashier $200 what will the change be." He told JS that in order to figure this out "I make it simple and by counting from 140 to 200 instead of subtracting. I count by tens: 150, 160, 170, 180, 190, 200 and the answer is $60."

            10   20 30   40   50 60

Based on the counting strategies that children use naturally, teachers can help them develop more efficient strategies, such as counting by tens, or hundreds, in settings where counting by ones would be too time consuming. That way children can extend the realm of what they can verify by themselves.

Children use particular examples to show that a statement is true

Many times children give one single example as evidence. Korri, 5th grade, age 10, stated "Multiplication reverses division and vice versa" She used the example 5×5 = 25, 25 ÷ 5 = 5 (CS). A child in the 3rd grade stated "You can't take away more than you have. When teacher gives me $5 dollars and I buy something, I have to stay less than that. If I don't, then I will not have enough money to get it." Another 3rd grader stated, "Adding and subtraction are the same just different directions" The student used the examples 10¢ + 5¢ = 15¢, and 15¢ - 5¢ = 10¢, to show that "this problem works both ways. Addition and subtraction are opposite" (BMe). Vanessa, 5th grade, stated "When you divide, the remainder cannot be the same or larger that the number you're dividing by." Example: "10 divided by 3 is 3 with a remainder of 1" (CT).

Some of the children give more than one example to show that the assertion is true. Anthony, 6th grade, stated: "When you divide a number by two you get half of the number you started with." He wrote the numbers 2, 4, and 6. He then divided each by two and got 1, 2, and 3 respectively. Then he did the opposite of that and multiplied each of the first numbers by two and got back to the original number (JDP). Girls in 2nd grade stated "When you add two numbers you get a larger number." As evidence they used the examples 1 + 1 = 2, 3 + 2 = 5, and 2 + 2 = 4 (EA).

Teachers can help students progress from giving just one example, to giving several examples, to giving examples in a systematic way and look for patterns. Patterns have the additional advantage of permitting children predict and then verify their prediction.

Children use mathematical relations to ascertain truth
In the case of some children, their explanations are rooted in a good grasp of the meaning of the mathematics they are doing. Olga, 7 years old, "Cuando le sumo el cero a un número es como no sumar nada. Cuando veo uno más cero, sé que es uno" [When I add zero to a number, it is the same as not adding anything. When I see one plus zero, I know it is one] (MEG). Robert, 4th grade, stated "When you times anything by zero, it always equals zero." He explained that "with the beans, zero groups of any number is zero" (JDP).

In some cases their explanations are based on methods they invent. Andrew, 4th grade, "Draw 6 lines from the equation 24 ÷ 6 =, draw a mark and count on every line up to the number to be divided (24) and the ... marks across equals [the answer]" (MF).

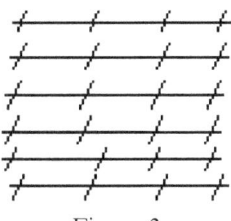

Figure 3

Children's explanations may also show tacit, and sometimes explicit and well articulated understanding of properties of numbers (like divisibility) and of operations (such as distributivity). For example, Isaiah's (age 8) explanation for $10 \times 11 = 110$ was "If $10 \times 10 = 100$, then one more ten added to that would be 110" (KP). Danielle, a fifth grader, explained how she knew that $9 \times 6 = 54$. She said she had a hard time with her nines times tables so she multiplies $10 \times 6 = 60$ and then takes away 6, equals 54. She said this works because 10 is one more than 9, so you have to subtract however many you are multiplying by (LM). Marisol, 5th grade, says that division is the opposite of multiplication. For example, to divide 81 by 9, Marisol keeps adding by 9 until she reaches 81, then she counts the number of times that she added (LM). Jesica, 2nd grade, stated "10, 20, 30, 40, 50, 60, ..., so on, are multiples of 10." She explained that "$10 + 10 = 20, 20 + 10 = 30, 30 + 10 = 40, 40 + 10 = 50, 50 + 10 = 60$, and that by adding 10 to every number you will get the next number in order, and that adding or subtracting 10, you will still end up with a multiple of 10" (AH).

Teachers can help students make explicit, and bring to the conscious fore the mathematical properties used by children. The technical vocabulary (such as distributive property) is not what is important, but realizing how such a property was used.

Children know it's true because someone they trust told them
Students learn many things from the people that are close to them, their parents and siblings. They trust them and believe what they learn from them. Karen, second grade, explained that tables were $1 + 1 = 2, 2 + 1 = 3, 3 + 1 = 4, ...$ TA asked her how she knew that this was true. She told TA that her dad taught her to memorize them because it was very important. TA asked her to prove that it was true that $3 + 1 = 4$. Karen kept writing $3 + 1 = 4$ and telling TA that it was true; and that TA just couldn't understand. Karen was getting very frustrated. She just wanted TA to believe her and didn't know how to prove it. Just being questioned about the facts she had learned from her father made her uncomfortable. As TA states, "She felt that I didn't believe her father." Then TA told her to prove it without writing it. Karen got some blocks in different shapes and tried to show TA by making a story. After some groping, Karen decided that she had 3 shapes and then put 1 more, with that she had 4 shapes. Gloria, 6th grade, knows that a negative times a negative equals a positive. She could not tell why, but she said that her older brother told her that that was true, so she knows that this is true (JDP).

We mentioned in the beginning that children also have a deep faith that their teachers are teaching them "the right thing." If Vincent, 3rd grade has a problem such as $20 + 3$, he knows how to find the answer by covering up the zero, which leaves the 2 and 3 exposed so he knows the answer is 23.

When BMi asked him how he knew this was correct, he said his teacher told him it was true. 4th grade students stated that 9 is odd because it goes odd, even, odd, even ... They could not really give a reason why it goes odd, even, odd, even. They said that is how the teacher always does it (JS).

Teachers can help students by letting children share their own explanations of things that they have learned. Children will realize that they can verify on their own the things that they have learned in mathematics from people they trust, and that it is an important aspect of mathematics to question why things work.

Children are unable or unaccustomed to explain why something is true

A common finding in these interviews was that it was hard for students to explain why the facts they had learned were true. After interviewing a group of fifth graders, and a group of fourth graders, JS wrote "With both sets of students they all knew a lot of facts but for about half of the problems they could not tell me the reason why or give me proof. They kept telling me they just knew it." After interviewing Olga (7 years old), MEG wrote "When I was listening to her sometimes she wouldn't know how to explain herself. CH wrote after her interviews

"The students had a really hard time explaining the math facts that they knew. In each case, both kids' first reaction was to say "I don't know" or "because that's what I was taught." Even after thinking a little, they still had trouble explaining what the math meant."

At times some students did provide an explanation that did not really prove anything. Vanessa, 5th grade, stated that "When you estimate, 5 and above you round up, and 4 and below you round down" and her explanation was "It's just because." (CT). Jesica, 2nd grade, stated "A triangle has three equal angles." When asked to prove it, all Jesica came up was a drawing of an equilateral triangle (AH). Also, not all children have clear understanding of what true means in mathematics. Melissa, 4th grade, told the interviewer that addition was true. Elaborating on this, "addition would always give you an answer" (MF).

In some cases the lack of ability to give a convincing argument may just be due to the fact that students are not used to give explanations, in other cases it may reflect also lack of understanding. Some students were able to explain after some initial confusion, other students were not. KR asked Katy, fifth grade, several multiplication questions, for example, what is five times three, and she rattled off the answers from memory. When KR asked her to show how she came up with the answers she was confused. KR then asked her to pretend that KR knew nothing about math and then Katy began to understand. Katy showed KR how, if you want to do three times two, you take three objects, in this case beans, three in each row and then counted all the beans to get six. Matthew, 4th grade, was asked the same multiplication questions but he was unable to show how he came up with his answers.

It was also obvious that many of the students had not much practice explaining their thinking in mathematics, nor giving convincing arguments or evidence to show that something they had learned in mathematics was in fact true. As teachers, we can provide opportunities for our students to explain and justify their answers, questioning not only when there is a mistake, but also when the result is correct or the statement true.

Final comments

The interviews were an eye opening experience for several reasons. One was the diverse methods exhibited by children. As one of the interviewers wrote This demonstrates a wide variety of how children make sense out of math. Some of their methods were very surprising to me and it made me look at math in a way I have never looked at it before. It is amazing to see the thought process of how a child's mind actually works and the things they can discover. (LM)

Another finding was that many children were uneasy about someone questioning the things they had learned in mathematics. Many students were baffled by the question "How do you know it is true?" The answers and attitudes of many of the children interviewed reveal a deep trust in what their parents, elder siblings, teachers, and books teach them.

In a few cases, some of the answers children gave were wrong, and they did not realize it. Cases we found included when children did not remember exactly how a number was defined (an 8 year old child talking about the relation between a googol and a googolplex), confusing one definition with another (for example, mode and arithmetic average), or over-generalizing a property of multiplication to division. By helping students develop ways to verify a mathematical statement some of these mistakes can be averted. In other cases, of course, the child will need outside information, for example, to look up a definition.

In other examples, children do not have the necessary understanding of how certain terms, that are also used in everyday life with different meaning, are used in mathematics, to ascertain whether a statement is true (for example kindergartners discussing how many sides does a circle or a square have).

On the other hand, many students showed self-reliance and clear understandings of the facts they had learned, many times using their own methods to show why the facts were true. It is on these understandings and methods that teachers can build to develop more efficient and systematic ways to provide evidence, and to help students grow in their mathematical thinking to include in their convincing arguments and proofs, reasoning of mathematical nature, of course, at the level of rigor appropriate for them.

One thing that was salient from the answers, is that many times students in upper grades still relied on methods which were very effective to deal with smaller numbers, but that their methods had not evolved, and were inefficient in more complex situations. Also it was clear that many of them had not developed a more sophisticated network of mathematical relations to deal with the more advanced ideas. Many of the students in the upper elementary or middle grades recalled only facts they had learned several years earlier. Most of the successful explanations were about facts with whole numbers. As children progress through more complex mathematics (fractions, negative numbers) their ability to explain why the facts they learn are true seems to lag behind. As teachers we need to help students make the transition from the empirical methods that work so well with small numbers to methods based on relationships between numbers and between operations.

To gain insight into the children's thinking, we teachers must ask them to explain or to justify what they are doing. We will certainly find a plethora of other ways that were not illustrated in this paper for lack of space (for example, how children use patterns to convince themselves). By understanding how children know it's true, teachers can better help students develop their natural ways to verify and convince themselves and others into the kind of reasoning that will eventually lead to mathematical proof in the higher grades.

Reference

Sowder, Larry, and Guershon Harel. Types of students' justifications. *Mathematics Teacher* 91 (November 1998): 670-675.

18 A RHYTHMIC APPROACH TO GEOMETRY[18]

Introduction

String designs have been used by several authors to help students develop and understand mathematical ideas. There are different aspects that can be stressed. First, there is the beauty of the final string design, in many cases related to the mirror and rotational symmetries involved (Pohl 1986). Second, string designs can also be used to highlight relations among numbers such as common factors, multiples, prime numbers (Bennett 1981; Perl 1981). String designs can also give rise to interesting mathematical curves (Millington 1996), such as parabolas, pursuit curves, spirals, cardioids, and many more.

Somervell (1975), in her classic book "A rhythmic approach to mathematics" proposed that young children's use string designs, and emphasized the dynamic and kinesthetic aspect of doing the designs to generate curves. According to her, children at a young age can get acquainted and develop pleasurable associations with these curves, so that later the mathematical treatment is seen by them as "orderly explanation of experiences long familiar" (Somervell 1975, p. 17).

In this article we will describe how a dynamical geometry program can be used in the same spirit by students in the middle grades. Interactive figures are available on line (Flores 2000). To better feel the rhythm of the creation of the designs students can interact on their own with them or they can watch the demonstration by the teacher. These figures were developed using JavaSketchpad (Jackiw 1998), the world wide web component of the Geometer's Sketchpad (Jackiw 1995). However, to use the interactive figures no previous knowledge of the software is needed. A brief set of instructions

---

[18] Flores, A. (2002). A rhythmic approach to geometry. *Mathematics Teaching in the Middle School, 7*, 378-383. Copyright National Council of Teachers of Mathematics. Used by permission.

is given in the introduction to this site (see figure 1). Whenever readers are referred to an interactive figure in the text, they can find it on this site. On the other hand, teachers and students familiar with the Geometer's Sketchpad or other dynamical geometry programs should not have problem to develop their own sketches following the descriptions given here.

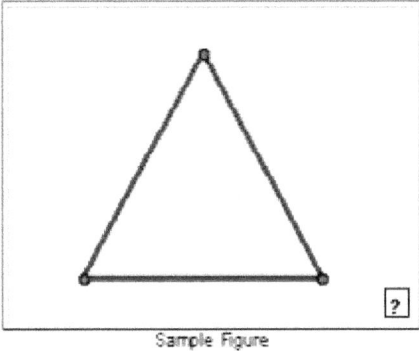

Sample Figure

Instructions

- Click and drag on red points to move them.
- Click on buttons (when displayed) inside the sketch to cause objects to move, or new objects to appear and disappear. (If a button stays down when you click it, you can click a second time to release it.)
- When objects are animating or moving, press '>' to speed up motion or '<' to slow it down.
- Press 'r' on your keyboard to reset the sketch to its initial configuration.
- If the screen becomes cluttered with traced objects, click the red 'X' (when displayed) in the lower-right corner to clear the traces.

Figure 1. Instructions for interactive figures on the web.

The basic idea of the rhythmic approach described here is that curves are obtained from straight segments that connect two moving points. As one of the points changes its position along a given path, the other point changes position along another path (or the same path but starting someplace else, or with a different speed). Geometer's Sketchapd has the capability to trace the succession of segments joining the moving points; the family of segments will form the desired curves. We will provide examples of curves such as parabolas, pursuit curves, spirals, a curve that resembles a caustic, then a real caustic (the cardioid), and the nephroid. Because this article is meant as an informal introduction to these curves at the middle grades, no emphasis on the abstract properties of these curves is made. Readers interested in the mathematical properties of these curves can consult Lockwood (1976) or Yates (1974).

The parabola

The parabola is obtained in the following way. Let AB and CD be two segments of equal length, point F is on AB, E on CD, and the segment FE joins the two points (see figure 2a). Point F moves with uniform speed starting at A. At the same time point E moves with the same speed starting at D (that is, it will move in "opposite" direction) (figure 2b). The trace feature of the Sketchpad will show how the segment FE changes as the points move (see figure 3a or interactive figure 2). The reader can experiment with different angles between the two lines. As you move point A up and down on interactive figure 3, what happens to the set of lines? What happens if you make the vertices coincide, or have them more separated? How do you need to manipulate the segments to obtain the shapes shown in figures 3b, 3c, and 3d? Where is the axis of symmetry of the parabola in each case?

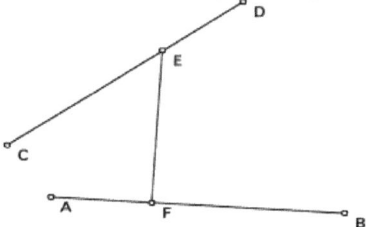

Figure 2a. The moving segment EF.

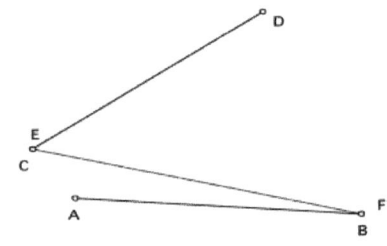

Figure 2b. Starting place of E and F.

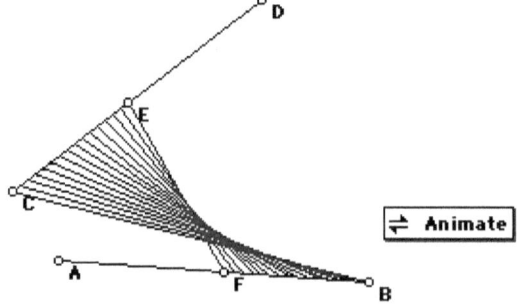

Figure 3a The parabola is being formed

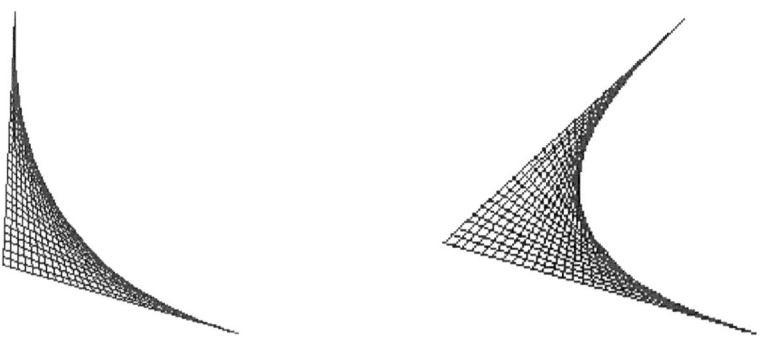

Figure 3b. Wide angle parabola　　　　Figure 3c. Acute angle parabola

Figure 3d. The fixed segments do not have to share a point

Multiple parabolas give rise to beautiful designs. With the Geometer's Sketchpad, students can animate several points along several paths. The purpose is not that student copy these designs, but to give examples of the kind of designs that students themselves will be able to discover. Even very young children are able to invent these kinds of designs with cards, needle, and string (Somervell 1975). In figure 4 (see also interactive figure 4), the points move from the vertices to the midpoints of the sides. In figure 5 (see also interactive figure 5) the points move along the lines from the vertices of the triangle to its center. In figure 6 the parabolas of different colors are generated by points moving on the sides of the triangle. (On interactive figure 6, remember to press the key 'r' to return the points to the initial position before animating again.) In figures 7 and 8 points move along the sides of a square (see interactive figures 7 and 8).

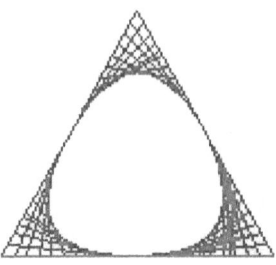 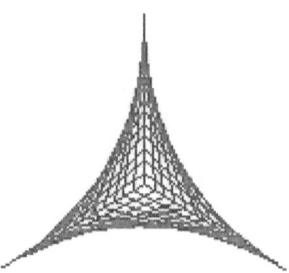

Figure 4. Parabolas on the sides   Figure 5. Parabolas on the medians

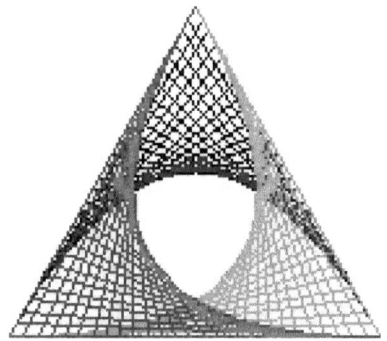

Figure 6. Three parabolas of different colors

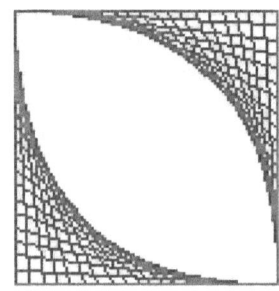   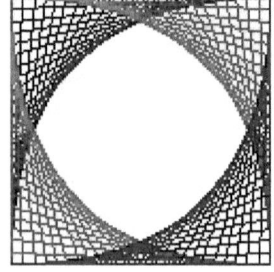

Figure 7. Two parabolas   Figure 8. Two pairs of parabolas

Pursuit curves

Pursuit curves receive their name because that would be the path of a predator chasing a prey that moves on a straight line, if the predator aims always towards the prey. A predator located at point C is trying to chase the prey located initially at point A. The red line shows where the predators is aiming (see figure 9a). When the prey sees that it is being chased, it runs directly towards its refuge at point B. Because the target is moving, the direction of the predator changes accordingly. After one step moving towards

A, the predator realizes that the prey has moved along segment AB, so the predator aims at the new position of the prey (see figure 9b). The lines pointing to the successive positions of the prey from the successive positions of the predator form a series of tangents to a curve of pursuit (see figure 10a, or interactive figure 10).

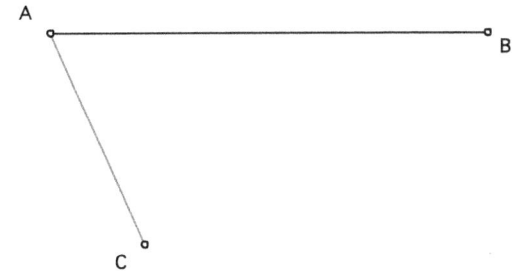

Figure 9a. Aiming at the prey, initial position.

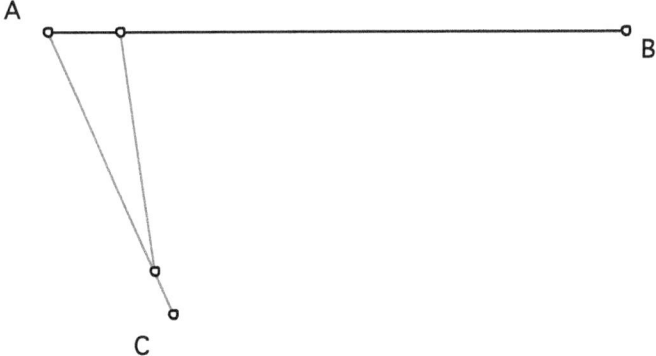

Figure 9b. New aim at the prey, after one step.

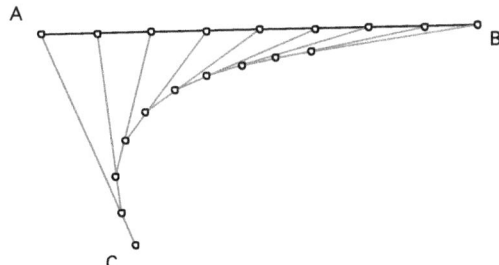

Figure 10a. The changing direction of the predator

Students can experiment with different initial positions of the predator and the prey, and with different relative angles. How do you need to position the initial points A, B and C to obtain the shapes shown in figures 10b, 10c, 10d? In the cases represented here, the prey reaches its refuge safely. The reader can change the relative speed of the predator on the interactive figure 10 by making segment DE bigger, and see how the pursuit curve changes.

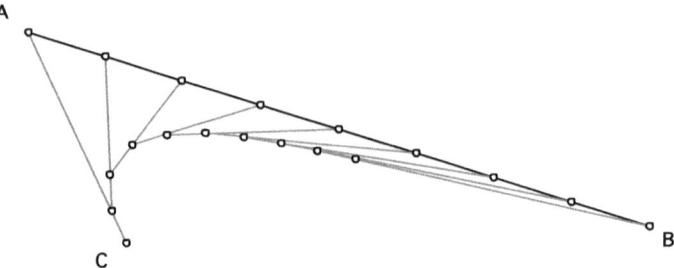

Figure 10b Slow pursuit

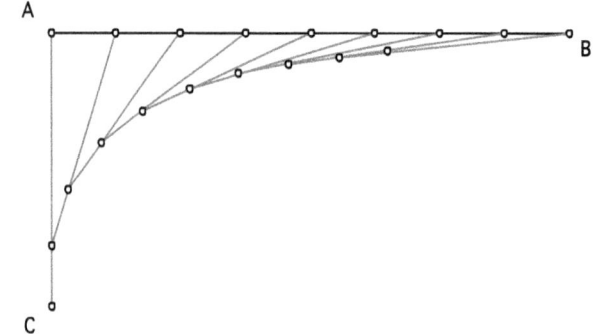

Figure 10c. Pursuit (another angle)

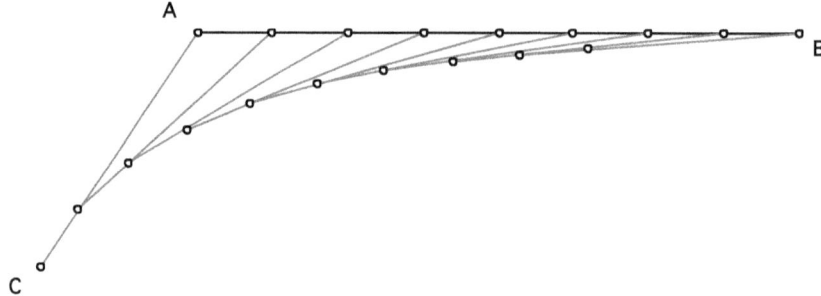

Fig 10d. Pursuit (obtuse angle)

Archimedian spirals

One kind of spirals can be obtained in the following way. Point G moves out away from D at constant speed along a line that is itself rotating at constant angular speed around the fixed point D (see figure 11). As the point moves away from the center, the line carries it around. The path of the point is a spiral, first studied by Archimedes (see figure 11a, or interactive figure 11). Students can experiment with different angular speeds, and different speeds along the line. In figure 11b the speed along the line has been increased, so that the point is farther away from the center after only a fraction of a turn. In figure 11c the speed along the line has been reduced, and the angular velocity increased, so that the point goes around the center several times, but will still not be very far away.

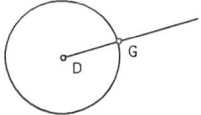

Figure 11. Point G moves away along a rotating line

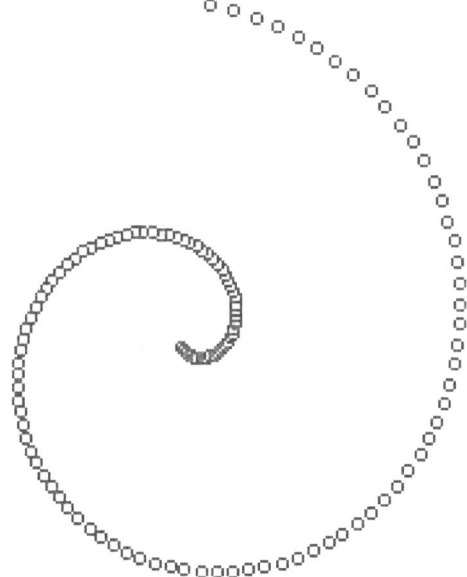

Figure 11a. Archimedean spiral

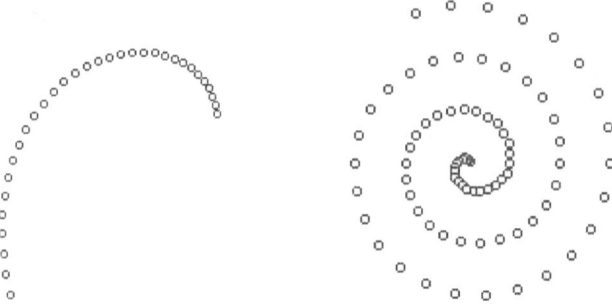

Figure 11b. Going out fast   Figure 11c. Going out slowly, rotating rapidly

Segments between two circles

An interesting design is obtained by the segments that connect points moving on concentric circles, one with twice the radius as the other. Point E moves around the outer circle with constant speed, as point F moves around the inner circle also with the same speed, so that the outer point circles once as the inner point circles twice. Segment EF joins the two moving points (see figure 12). As the two points move, segment EF traces a curve (see figure 13, or interactive figure 13).

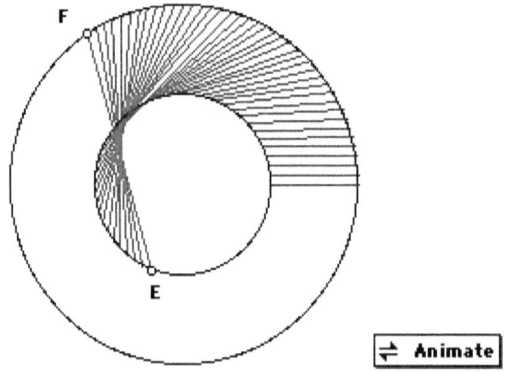

Figure 12. Point E and F move with the same speed.

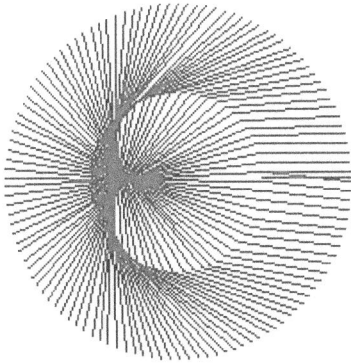

Figure 13. Lines between two circles.

Cardioid

The reader may have glimpsed a cardioid, the curve of light rays that is formed on the bottom of a cylindrical cup when light is reflected on its walls. We can generate a cardioid (which means in the shape of a heart) by having two points on the same circle, both moving in the same direction; both start at the same place, but one is going twice as fast as the other. If we connect the two points by a segment and trace the different positions of the segments as the points move (see figure 14a) we generate a cardioid (see figure 14b, or interactive figure 14). The resemblance between the curve obtained in the previous example and the cardioid is only superficial.

Point G can be animated to go twice as fast as point as point F in the following way. F is animated around a circle. D is animated around a smaller circle (hidden), concentric to the first, and with a radius that is half as big as the radius of the first circle (see figure 15). G is the projection of D on the bigger circle. Because the speed of D is the same as the speed of F, but the radius of its path half as small, its angular speed (and therefore that of G) will be twice as big as that of F.

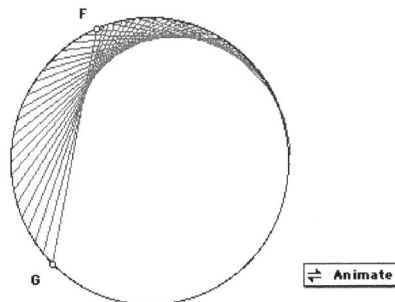

Figure 14a. G moves twice as fast as F.

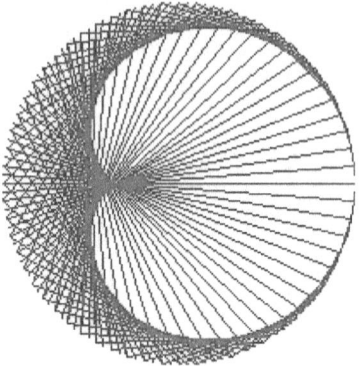

Figure 14b Cardioid

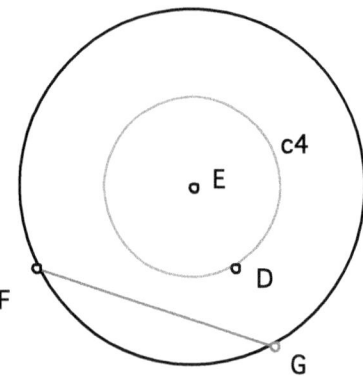

Figure 15. Points with different speeds on the same circle.

Nephroid

In a similar situation of two points moving around the same circle, by changing the relative speed of the two points, students can generate other curves. When one point moves three times faster than the other, students can obtain a curve called the nephroid ("kidney-shaped"). See figure 16 or interactive figure 16.

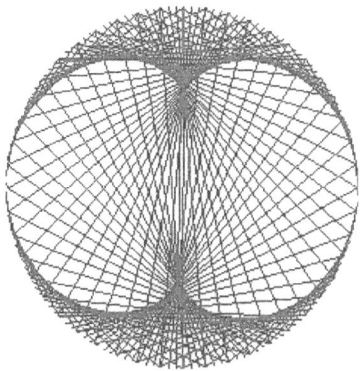

Figure 16. Nephroid

In addition to families of line segments, families of circles can form also curves. The nephroid can also be formed by the family of circles that have their center on a base circle and are tangent to a fixed diameter of that circle (see figure 17). As the moving point goes around the fixed circle, circles of different sizes are generated (see figure 18 or interactive figure 18).

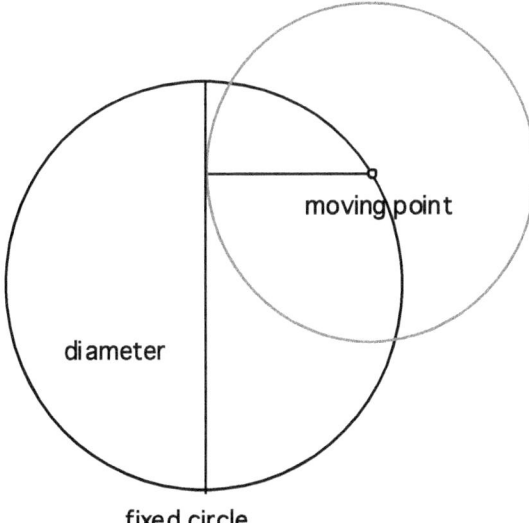

Figure 17. Red circle has center on a circle and is tangent to the diameter.

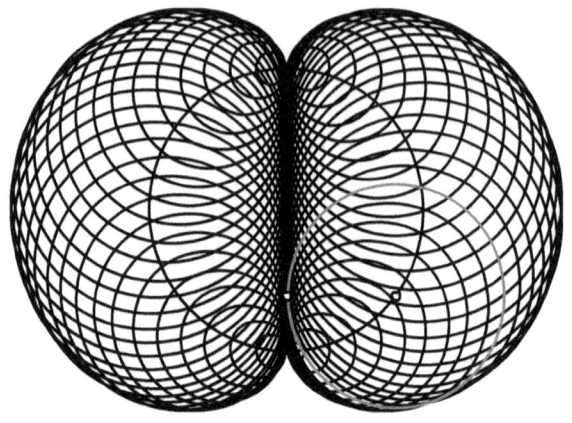

Figure 18 Nephroid as envelope of circles.

Conclusion

All too often, students' first approach to mathematical objects is via an abstract representation such as a formula or an equation. In the case of curves, many times students have to deal with their analytic definitions before they develop familiarity with many examples and develop an intuitive grasp about them. The rhythmic approach to geometry, using a dynamical geometry program can build on previous or simultaneous experiences with string designs and permit students in the middle grades develop familiarity with curves through a kinesthetic experience. At the same time, students can experiment and enjoy the inherent beauty of the designs. Beauty in mathematics is an aspect not emphasized enough in the middle grades.

References

Bennett, Albert B. Star patterns. In Kenneth E. Easterday, Loren L. Henry, and F. Morgan Simpson (Comps.) *Activities for junior high school and middle school mathematics* (p. 48-50). Reston, Va.: National Council of Teachers of Mathematics, 1981.

Flores, Alfinio. *Figures for "A Rhythmic approach to geometry"*. [On line]. Available at http://www.public.asu.edu/~aaafp/rhythm.html, 2000.

Jackiw, Nicholas. *The Geometer's Sketchpad 3.0* (Computer software). Berkeley, Ca.: Key Curriculum Press, 1995.

Jackiw, Nicholas. *JavaSketchpad* [On line]. Available http://www.keypress.com/sketchpad/java_gsp/index.html, 1998.

Lockwood, E. H. *A book of curves*. Cambridge, Eng.: Cambridge University Press, 1976.

Millington, Jon. *Curve stitching.* Norfolk, England: Tarquin Publications, 1996.

Perl, Teri. String sculpture in the mathematics laboratory. In Kenneth E. Easterday, Loren L. Henry, and F. Morgan Simpson (Comps.) *Activities for junior high school and middle school mathematics* (p. 43-47). Reston, Va.: National Council of Teachers of Mathematics, 1981.

Pohl, Victoria. *How to enrich geometry using string designs.* Reston, Va.: National Council of Teachers of Mathematics, 1986.

Somervell, Edith L. *A rhythmic approach to mathematics.* Reston, Va.: National Council of Teachers of Mathematics, 1975.

Yates, R. C. *Curves and their properties.* Reston, Va.: National Council of Teachers of Mathematics, 1974.

19 WHY DON'T TEACHERS KNOW ALL THE WAYS?[19]

Abstract

This article discusses how important it is for teachers to be open to alternative procedures and strategies students may use. A method that is meaningful rather than one learned by rote will help students develop more confidence and understanding.

In order to help all students become confident in mathematics, teachers need to know and accept a variety of ways to tackle a problem. This was keenly made explicit by a student named Steven. Due to an unstable familial situation, he has attended a different school nearly every year. At nearly every move, he was told that the way he learned mathematics at his previous school was "wrong". The *right* way was the way presented by his current teacher. Steven's attempt to keep straight all the various algorithms he was taught along the way has resulted in a confused and very frustrated boy. He would often hide behind a facade of "superior knowledge"; as he was older than the other boys he would act superior or arrogant.

The good news is that he has never given up. There are some positive changes in his life too. Steven is now in a stable environment as he lives with his grandparents. In school, he is now a student in a class where the teacher developed an environment that encourages students to ask questions. He has finally built up the confidence to ask, "I learned it like this. Is it OK if I do it

---

like this? I don't understand your way."

He was very surprised when the teacher told him, "Yeah, that is another way to do it. There are lots of ways to do problems. Yours is equivalent to mine. Sure." Steven almost fell off his chair and exclaimed: "Why don't teachers know all the ways?"

It is natural for teachers to teach the method they learned. It is also common to insist on using the method prescribed and not deviate from established practice in a school. Often this is due to lack of confidence in mathematics on the part of the teacher and fear to be embarrassed. It can be intimidating for a teacher not to be able to understand a student's method.

Feynman, an outstanding physicist and one of the sharpest scientific minds of the 20th century, relates a story that illustrates that we use the method we are taught and we cling to that way, and no other way will do. In the story, Feynman is not quite a teenager; his cousin, who is three years older, is having problems in algebra. Feynman asks his cousin what he is trying to do. The cousin explains that he is dealing with a problem where you know that $2x + 7$ equals 15 and you are trying to find x. "You mean 4" says Feynman. The cousin retorts "Yeah, but you did it with arithmetic, you have to do it by algebra" (Feynman 2000, p. 6). Feynman points out that the cousin was never able to do algebra, because he did not understand how he was supposed to do it.

When Feynman learned algebra on his own, it did not make any difference how he did it. In contrast, he observes that schools often teach students to follow a set of rules that can be followed without thinking and produce an answer. Often, students are taught so many rules and exceptions, and they have the idea that the way the teacher taught it is the one and only way, even if they do not fully understand it. As Feynman states, when people don't learn by understanding their knowledge is so fragile (1985, p. 36-37). They flounder when they see something new and think that it is altogether different.

Students can develop confidence in learning mathematics when they are allowed to use methods they understand, even if it is not the method prescribed in that chapter. Sometimes they use methods they learned someplace else. Often, they invent their own methods. Also, allowing students to share other ways helps all learners. Some of the other students like these alternative ways and are pleasantly surprised to hear that the teacher will accept them. Teachers need to be open to students methods for solving problems. By doing so they also encourage students to take control of their own learning. This active role of the students is clearly in consonance with what we know about how they learn best (NCTM 1991, 2000).

Of course, it would be impossible for teachers to know all the methods. Rather, they need to be willing to accept students' methods, try to understand how and why they work, and how these alternative methods are related to the other methods used in class. In many cases, this requires from teachers to become better mathematicians themselves. Teachers can become more confident in their abilities as mathematicians through course work and workshops, by reading professional journals, and by sharing with other teachers and participating in professional organizations.

References

Richard P. Feynman. *"Surely you're joking, Mr. Feynman!"* New York: Norton, 1985.

Feynman, Richard. *The pleasure of finding things out.* Cambridge, Mass.: Perseus Publishing, 2000.

National Council of Teachers of Mathematics. *Professional Standards for Teaching Mathematics.* Reston, Va.: NCTM, 1991.

National Council of Teachers of Mathematics. *Principles and Standards for School Mathematics.* Reston, Va.: NCTM, 2000.

20 MATHEMATICAL NOTATIONS AND PROCEDURES OF RECENT IMMIGRANT STUDENTS[20]

Introduction

Mathematics is oftentimes referred to as a universal language. Compared to the differences in language and culture that recent immigrants face when they arrive to the United States, the differences in mathematical notation and procedures seem to be minor. Nevertheless, there are noticeable differences between the way mathematical ideas are represented in their country of origin and in the United States. If not addressed, the differences in notation and procedures can add to the difficulties that immigrants face during their first years in their new country.

Exposing teachers to these differences expands their repertoire and gives them an appreciation of their student's previous experiences and struggles. For example, the statement $59 : 8 = 7 + 3 : 8$ might cause a teacher in the United States to pause until he realizes that $:$ can also denote division; the statement is another way to express the conversion of $\frac{59}{8}$ to $7\frac{3}{8}$. In the same way, an immigrant student might hesitate when facing an unfamiliar notation. Also, confusion may arise when parents who were schooled in other countries try to help their children using procedures that are different from those taught at school (Ron, 1998).

Teachers are often unaware that an immigrant student is confused or has a question because the student does not always speak up. Many students come

[20] Perkins, I. and Flores, A. (2002). Mathematical notations and procedures of recent immigrant students. *Mathematics Teaching in the Middle School*, 7, 346-351.

from a tradition that one does not question the teacher, and it often takes several weeks before the student is confident enough to ask a question. If the teacher knows other algorithms, he can better include immigrant students by incorporating the other algorithm with the preface, "in other countries you might see this problem done in the following way."

As teachers encounter alternative algorithms taught in other countries, they realize that the algorithms they have learned are just one of the possible ways to compute the answer. This can help them become more accepting when students deviate from the procedure or algorithm taught in class and use their own procedures.

In this article, we will describe some differences in representations of mathematical concepts and procedures that recent immigrants from Latin America face in schools in the United Schools. We do not attempt to provide a description of all the differences that immigrants face in school mathematics. Rather, the purpose is to help students and teachers be aware of those differences and use them to the advantage of the students.

Most of the differences reported here were collected through direct observations by the first author in her classrooms consisting of recent immigrants during a period of three years. The second author, schooled in Mexico, provided some additional examples. We will discuss first notational differences. Then we will discuss algorithmic differences, and we finish the chapter with some considerations for success of recent immigrants in their new school system.

Notational Differences

Numbers. With respect to the symbols for numbers, the same Hindu Arabic system is used in Latin America as well as in the United states. However, there are sometimes differences in the way numbers are written; in the names that numbers are read; in the use of the decimal point; and in the separation of figures in big numbers. We will briefly describe these.

There are differences in the manner that students write numerals. Immigrants students may put a cross hatch in their number 7 to distinguish it from their 1. That is usually not a problem for people in the United States, although some say that the seven looks like a handwritten F. However, it can be a problem for immigrants because they cannot always distinguish when someone from the United States writes a 1 or a 7.

Another difference is that often immigrant students begin their stroke at the lower part of the numeral, while people in the United States begin their strokes at the top of the completed numeral. The completed numerals have a different "flavor" from those made by students in the United States. The most distinctive numerals are four and eight. Problems may occurs when a student curves the final upstroke of his 4 because he is rushed or nervous.

The result is that the finished numeral often looks like a 9 to both students and teachers. It is evident when grading papers or examinations that students who clearly understand the concepts make operational errors caused by this. Other differences occur in students from other countries. For example, recent immigrants from Cuba write their number 9 in a way that closely resembles the letter "g". The difficulty here lies with the teacher who might not be familiar with this form. See table 1 for a summary of differences in handwritten numerals.

Table 1.
Differences in writing numerals by hand

United States	Latin America	Comments
7	/ 7 7	The numeral 7; includes a cross hatched line. Sometimes it is drawn from the bottom up.
8	/ 9 8	The numeral 8; it is often drawn from the bottom up.
4	/ 4 4	The numeral 4; sometimes drawn from the bottom up. There is a tendency for students to confuse their 4's and 9's.
9	g	The numeral 9; particularly among the Cuban students.

There is a difference in the manner in which numbers are read. Both systems are identical through millions; they do differ, however, at the level which books in the United States identify as billions. Both a student in the United States and one from Latin America will read the number 782,621,751 as 782 million, 621 thousand 751. There is a difference when the number becomes 10,782,621,751. A student schooled in the United States will say 10 billion 782 million, 621 thousand 751. Some students who in their country of origin

213

learned to read such a number as "10 mil 782 *millones* 621 mil 751", read it in English as 10 thousand 782 million 621 thousand 751. A student from Latin America (and other countries like Great Britain) will not designate billions until there are at least 13 digits in a number. The number 23,500,000,000,000 is read by a student in the United States as 23 trillion 500 billion, and by a Latin American as 23 billion 500 thousand million (Serralde, Zúñiga, Zúñiga, & Zúñiga, 1993; Sperling & Levinson, 1988).

In the United States people separate numbers into multiples of thousands by commas. In some countries of Latin America they use the point to designate multiples of thousands. Consequently, the numbers 10,752,101 is equivalent to the number 10.752.101 (Secada, 1983). Other textbooks (Serralde et al., 1993) leave a space between multiples of thousands and write 10 753 101. In Mexico a third manner is used (Secretaría de Educación Pública, 1993a, 1993b). Millions are separated by an apostrophe and multiples of thousands are separated by a commas, as in 10'752,101. The semicolon ; is also used in Mexico to separate millions from thousands (Salas Luna, Jardines, & Ramones, 1997). For example, the surface area of Mexico (in km2) is written as 1;958,201.

In Mexico, negative numbers can be expressed in either of two ways. One is with a preceding minus sign (-2) or with the sign over the number ($\bar{2}$). Some students who are used to this last notation have problems with the notation for repeating decimal fractions used in the United States. The overhead bar indicates a decimal fraction which repeats. For example, the repeating decimal 0.3333333… is designated as $0.\bar{3}$. Some Mexican texts indicate a repeating decimal with an overhead arc, for example, .$\hat{3}$ (Beristáin & Campos, 1993).

Measurement. Immigrants find the units of measures used in the United States confusing. They are accustomed to the metric system in which all measures are interrelated. Divisions within the metric system are based on powers of ten, while the English system does not have consistent subdivisions, for example, 12 inches to the foot, and three feet to the yard. Linear measures in the metric system include the meter which is divided by ten into 10 decimeters; by a hundred into 100 centimeters; by a thousand into 1000 millimeters. The meter can be multiplied by 1000 to form the kilometer. Sometimes immigrant students from Latin America find in textbooks in the United States linear metric measures that are seldom used in their countries of origin (like the decameter and the hectometer).

Other symbols. A symbol which is not often seen in texts in the United States is ≅, which means "is approximately equal to" (Beristáin & Campos, 1993). The counterpart in the United States is ≈ .

Sometimes symbols are identical but differ in their positioning. Both in the

United States and Mexico textbooks use the symbol + to indicate addition. However, they differ in the position of the + with regard to the written operation (Serralde et al., 1993).

United States   Mexico

```
    45                    45
     9                 +   9
 + 128                   128
```

Students often set up addition problems in this manner, but there is no confusion on either part of teacher and students.

Differences in placement of the point creates several problems. The point is used in two very different ways in the texts in the United States. First, it can be used to designate decimal fractions. The number 2.54 refers to 2 whole parts and 54 hundredths of another. In this case the point is at the very bottom and between the numbers in question. Some countries in Latin America use the comma for that function. The decimal number 2.54 would there be written as 2,54 (Secada, 1983). In the United States the point is also used to indicate multiplication. In that case, it is placed between two numbers but half way up and between the given numbers; 2 · 54 refers to 2 times 54. A thicker point is also used sometimes in Mexico to indicate multiplication; the notation 2 • 54 indicates the product of 2 and 54 (Almaguer, Bazaldúa, Cantú, & Rodríguez, 1994). In other countries, the dot used to multiply is placed at the lower part and between the given numbers; the notation 2 . 54 in this case indicates the product of 2 and 54 and not a decimal fraction. Due to these differences, a student may interpret the notation 3.789 in a variety of ways. He may assume that the notation refers to the number three thousand, seven hundred eighty nine. In this case he may delete the point the belief that it is merely a "spacer" between thousands. Second, he may assume that he is being asked to multiply 3 and 789. Third, he may assume that the notation is the decimal expression of 3 whole parts and 789 thousandths of another.

Four different symbols are used to denote division in Mexican texts, namely $\div$ / : $\overline{)}$. All symbols are used in the United States, but there may be some confusion with the use of the colon. In the United States the colon is seen primarily in ratios and proportions, but in Latin America, it is also used to designate division. The division of the fractions $\frac{-3}{4}$ and $\frac{-3}{5}$ might be written as $\frac{-3}{4} : \frac{-3}{5}$; the division of 16 by 2 can be written as 16 : 2 = 8. Similarly the equation 4 : 5 = 8 : x has the solution $x = 10$ (Serralde et al., 1993; Beristáin & Campos, 1993).

Another notation difference is that used in angles. Texts in the United States might note angles in one of several ways:

$\angle \alpha$ $\angle ABC$ $\angle 1$

Some Mexican texts write the same angles as

$\overset{\wedge}{\alpha}$ $\overset{\frown}{ABC}$ $\overset{\wedge}{1}$

The problem is that students try to write the angle symbol as the teacher directs, but they make the symbol in the manner they were previously taught and place it on the side. The result often looks like this:

<α   < ABC   <1

The confusion is apparent when students begin to work with inequalities. The symbols look the same as angles. The question becomes why are we doing operations on angles? On the part of the teacher, the question becomes why are the students writing the angles like inequalities?

One needs to point out notational differences before the beginning of any lesson. It is important to validate the experiences of the students and to point out how knowing about differing systems enriches their knowledge. It establishes the idea that there are alternative ways to look at and represent mathematics.

Algorithmic Differences

The presentation of mathematical procedures may vary between countries. An immigrant student often enters a mathematics class with differing algorithms for a given operation. Frequently the teacher knows only the algorithms he was taught and is unaware of alternative methods.

Basic facts. It is not uncommon to see multiplication tables in the United States; however, students in Mexico have been taught arithmetic facts using not only multiplication tables but also addition tables. Immigrant students were often expected to have these tables memorized. The incorporation of these tables is the first step in computation. Students with interrupted schooling often say "I do not know my tables", when they are unable to remember an addition or multiplication fact. In order to help students with interrupted schooling "catch-up" in mathematics, a calculator can be used. However, sometimes there is opposition from students and parents. One tool against opposition are the guidelines from the Secretaría de Educación Pública (1993) which call for use of calculators. Also, the National Council of Mathematics Teachers (2000) calls for the availability of calculators to all students at all times.

In some places, intermediate steps in a procedure are computed mentally. Many recent immigrants take pride in being able to compute quickly and accurately in their heads. As a result, these students often merely write the answer to a given problem and omit the intermediate steps in their written work. Teachers unfamiliar with the students' backgrounds might wrongly assume the student has copied another student's work.

Addition and subtraction. Finger counting can be different. Frequently, Mexican students use their left thumb as the counter and the left fingers for the manipulation (if they write with the right hand). To add 29 and 3, a student says "29", and then he touches his left thumb to his left index finger and counts "30, 31, 32," as he moves to the little finger. He reverses the process to subtract. To subtract 2 from 30, he would say "30" and then touch the left thumb to the little finger, and move to the index as he counts "29, 28".

A subtraction algorithm taught in Mexico is based on the fact that the difference will not change if the same number is added to both subtrahend and minuend. For example, if 10 is added to both numbers, as ten units in the minuend, and as a unit in the tens place in the subtrahend, the difference remains the same (figure 1). In this algorithm there is no need to "borrow" from the column to the left when the number in a column on top is smaller than the number below. Rather, the student adds ten units to one column in the number on top and in the next step adds one unit to the column to the left of the number below to compensate (see table 2). The reaction of many teachers in the United States is that the student does subtraction backwards.

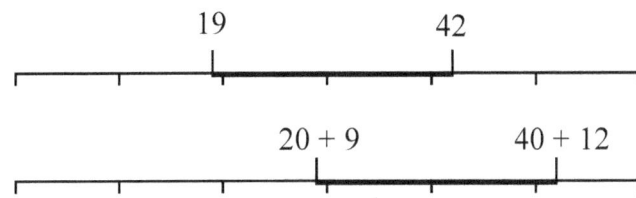

Figure 1. Constant difference

Table 2.
Subtraction algorithm based on missing addend

Written form	Thought process
542 - 269 3	9 to 12?, 3 (Notice that ten units were added mentally to the upper number to convert the 2 into a 12). Write the 3, and add 1 (ten) mentally to the 6 (tens). (Notice that we add one ten to the lower number in a different column).
542 −269 73	7 (tens) to 14 (tens)? 7 (tens), Write down the 7 (tens), and add mentally 1 (hundred) to the 2 (hundreds) in the next column.
542 -269 273	3 (hundreds) to 5 (hundreds)? 2 (hundreds), write down the 2 (hundreds), and we are done.

Rounding. In the United States, rounding is frequently taught by a set of rules. First, locate the digit in the place to which you are rounding. Increase this digit by 1 if the next digit to the right is 5 or greater; or leave the digit unchanged if the next digit to the right is less than five (Lynch & Olmstead, 1993). In Mexico, the number line is often used to round numbers (Beristáin & Campos, 1993). The values 600, 700 and 675 are plotted. Because 675 is closer to 700 than to 600, 675 is rounded to 700.

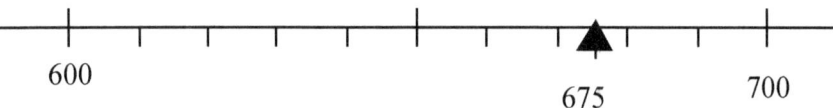

Figure 2. Rounding on the number line

Integer numbers. Number lines are also used to teach addition of integers in both countries. Lynch and Olmstead (1993) draw a number line and then use arrows to add two integers. The example in figure 3 demonstrates the equation 5 + (-4) = 1.

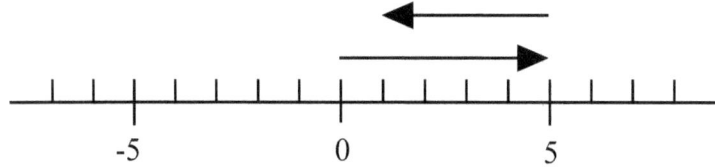

Figure 3. Adding 5 + (-4)

Beristáin and Campos (1993) use a pair of number lines printed on two cards which slide along each other (see figure 2). To add 5 + (-4), the lower card is positioned so that its 0 is below the other card's 5. Then one goes to the lower card (-4) and reads the resulting number, 1, on the upper card (see figure 4).

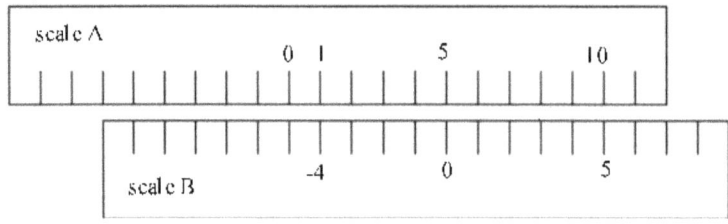

Figure 4. Adding 5 + (-4) with sliding cards.

Parentheses and distributive property. There are differences in approaches when multiplying a number by an expression in parenthesis. Students in the United States are taught to do all work within the parenthesis before any other operation. The expression

$2(3 + 5 - 2)$ is evaluated by doing $3 + 5 - 2$ first, and then multiplying the result, 6, by 2. In Mexican textbooks the expression is evaluated by using the distributive property,

$2 (3 + 5 - 2) = 2 \cdot 3 + 2 \cdot 5 - 2 \cdot 2 = 6 + 10 - 4 = 12$. The two procedures, of course, are mathematically equivalent, but to a young learner they can be confusing if the connection between the two is not made explicit.

Division. Division is done in similar manner but with the difference that more steps are written in the version taught in the United States (long division). In the Mexican version some steps are done mentally and only some of the intermediate steps are written down. Often, immigrant students take pride in being able to do intermediate steps mentally. Some of them consider long division for smaller children: "Nomás para los niños en la primaria, maestra". The long division algorithm for $126 \div 3$ is illustrated on the left column of table 3. In the shorter version, the subtractions are done mentally. The steps are illustrated in the middle column of table 3. Another way of writing the steps learned in other countries (Honduras and Cuba) is illustrated in the third column of table 3.

Table 3. Three ways to divide $126 \div 3$

Long division	Short division	Another form of short division
$\begin{array}{r} 42 \\ \hline 3)126 \\ -12 \\ \hline 06 \\ -6 \\ \hline 0 \end{array}$	$\begin{array}{r} 42 \\ \hline 3)126 \\ 06 \\ 0 \end{array}$	$\begin{array}{l} 126 \underline{)\ 3} \\ 06 \quad 42 \\ 0 \end{array}$

Prime factorization. Texts in the United States generally use a factor tree to systematically find prime factors. To find the factors in 140, it is divided into 2 and 70. Two is a prime number but 70 can be further divided into 2 and 35. The process continues until all numbers on the branches are prime numbers. However, one problem that arises is that some students often focus on the 5 and the 7 as "the answer" and forget about the two times that 2 appears as a factor.

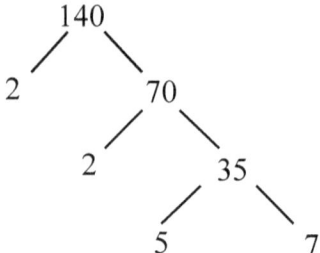

Figure 5. Prime factorization tree.

Mexican texts use a vertical line to accomplish the same process. The first prime is 2; 140 divided by 2 is 70; 70 divided by 2 is 35; 35 is not divisible by 2 nor by the next prime 3, therefore the next prime 5 is tried. The final prime factor is 7. All prime factors appear on the column to the right.

140	2
70	2
35	5
7	7
1	

Comparing fractions. Several approaches to comparing fractions are found in Mexican textbooks. To determine if a fraction is greater than, less than, or equal to another, some texts plot the fractions on a number line, others rewrite the fractions so that they have the same denominator, or use crossed products to compare the fractions. For example, $\frac{7}{50} < \frac{4}{25}$ because $\frac{7}{50} < \frac{8}{50}$; and $\frac{3}{5} < \frac{3}{4}$ because 3×4 < 3×5 (Salas Luna et al., 1997).

Finding common denominators. In texts in the United States, when one wants to change fractions to obtain same denominators, one generates a list of multiples for each denominator and then picks out the lowest common denominator. In Mexican textbooks both denominators are decomposed into primes. The lowest common denominator is found by multiplying all the common prime factors and the prime factors that appear in at least one of the two denominators.

4	6		2	(prime common factor)
2			2	(prime factor of 4)
1	3		3	(prime factor of 6)
	1			

The lowest common multiple of 4 and 6 is therefore $2 \times 2 \times 3 = 12$.

When adding fractions, a common denominator is found as part of the algorithm $\frac{a}{b} + \frac{c}{d} = \frac{ad+cb}{bd}$. To add fractions, first multiply denominators to obtain the new denominator; then cross multiply numerators and denominators and add the products; then simplify the result (Beristáin & Campos, 1993).

$$\frac{3}{8} + \frac{1}{4} = \frac{12 + 8}{32} = \frac{20}{32} = \frac{5}{8}$$

<u>Division of fractions</u>. The most common algorithm in the United States to divide fractions is to invert the second fraction and then multiply. In Mexico a common algorithm for a division like $\frac{1}{2} \div \frac{3}{4}$ is to cross multiply as shown in figure 6. Teachers can help students see how the two algorithms are equivalent beyond the mere symbolic manipulations. For example, both algorithms can be seen as shortcuts when fractions with the same denominators are divided (in which case all you need to do is divide numerators).

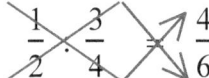

Figure 6. Division of fractions.

<u>Solving algebraic equations</u>. The equation $x + 35 = 75$ is solved in the United States by subtracting 35 from both sides so that $x = 40$. Mexican students write the original problem the same way, but the thought process is different. Their mental thinking is "What number and 35 are 75? 40 and 35 are 70." The written procedure shows

$x + 35 = 75$

$40 + 35 = 75$

$75 = 75$

Many times the student does not write $x = 40$ explicitly.

<u>Multiplication of binomials</u>. In Mexican texts, arrows are used in the multiplication of binomials and polynomials (Serralde et al., 1993). To multiply two binomials, the diagram indicates that the first term of the first parenthesis is multiplied by each of the terms in the second parenthesis; then the second term of the first parenthesis is multiplied by each of the terms in the second parenthesis (see figure 7), which gives the sum $9x^2 + 6x + 18x + 12$.

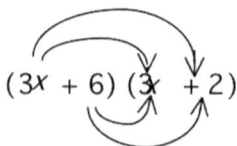

$(3x + 6) (3x + 2)$

Figure 7. Order for multiplication of terms

In the United States, textbooks often use the FOIL mnemonic (First, Outer, Inner, Last) which refers to the order in which the terms are multiplied. The FOIL mnemonic, of course, cannot be translated into Spanish. Although the two methods give the same intermediate results, the method used in Mexican texts can be readily extended to multiplication of polynomials with more than two terms, while the FOIL mnemonic cannot.

The algorithms presented in this section are not all the differences in procedures that immigrant students face in the United. They however establish the idea that there are various differences between countries, and show that even basic ideas can become confusing when presented through a different representation.

Considerations for success

The previous sections have identified situations which an immigrant from Latin America may find difficult. If the immigrant student is to be successful in a school in the United States, what can the classroom teacher do to help him or her? Here are some recommendations.

First, validate students' previous experiences both linguistically and mathematically. Emphasize the richness and diversity of their knowledge and of their experiences with multiple systems of knowledge and expression. Most have had schooling in their country of origin; that experience differs from their school experience in their new country. Highlight those experiences in a positive light and eliminate the notion that the ways of the schools in the United States are inherently better. Take time to explain differences students encounter. To do so creates a comfortable environment where students know they can express themselves without fear. When students know that if they do something a little different it will be accepted, they are more relaxed and confident.

Second, find common beginning points for student to start their experiences in the United States. One can begin with scientific or graphing calculators. Most immigrant students did not use calculators in their home schools. The use of written explanations is also a good common ground; students are not used to writing explanations in mathematics. In the beginning students may not be happy, but they soon will listen critically to the teacher's explanations and to raise questions if they felt they were unsatisfactory.

Finally, establish a sense of rapport in which both students and teachers are learners. For example, it can be the case that the Spanish of the teacher is not as good as that of the students. Students can help improve the teacher's Spanish as the teacher helps them master English. Allow class time for students to practice the variety of language used in mathematics, in both formal and in informal settings, both in English and in Spanish. Allow students to share their algorithms. Algorithms from different countries may differ. Sharing these further enriches the class as a whole.

Of course, in addition to differences in representations there are also differences in the ways mathematics topics are sequenced in different countries. The teacher can provide enrichment and challenging activities for a topic that, although new at a certain grade level in the United States, is known to students who come from other countries.

The teacher plays a pivotal role in the success of immigrant students. The key lies in the teacher making an active decision to do so.

References

Almaguer, G., Bazaldúa, J. M., Cantú, F., & Rodríguez, L. (1994). *Matemáticas 2*. Mexico City: Limusa.

Beristáin, E., & Campos, Y. (1993). *Matemátias y realidad 1*. Mexico City: Ediciones Pedagógicas.

Lynch, C., & Olmstead, E. (1993). *Mathmatters 1*. Cincinnati, OH: South-Western.

National Council of Mathematics Teachers. (2000). *Principles and standards for school mathematics*. Reston, VA: Author.

Ron, P. (1998). My family taught me this way. In L. J. Morrow & M. J. Kenney (Eds.), *The teaching and learning of algorithms in school mathematics* (pp. 115-119). Reston, VA: National Council of Mathematics Teachers.

Salas Luna, M. S., Jardines Garza, F. J., & Ramones Martínez, M. (1997). *Matemáticas 2: Educación secundaria segundo grado*. Monterrey, Nuevo León: Ediciones Castillo.

Secada, W. G. *The educational background of limited English proficient students: Implications for the arithmetic classroom*. Arlington Heights, IL: Bilingual Education Service Center, 1983. ERIC ED 237 318

Secretaría de Educación Pública. (1993a). *Educación básica primaria: Plan y programas de estudio*. Mexico City: Author.

Secretaría de Educación Pública. (1993b). *Educación básica secundaria: Plan y programas de estudio*. Mexico City: Author.

Serralde, E., Zúñiga, J., Zúñiga, H., & Zúñiga, E. (1993). Matemáticas Uno. Mexico City: Ediciones Pedagógicas.

Sperling, A. P., & Levinson. S. D. (1988). *Arithmetic made simple*. New York: Doubleday.

21 INTEGRATION OF TECHNOLOGY, SCIENCE, AND MATHEMATICS IN THE MIDDLE GRADES: A TEACHER PREPARATION PROGRAM[21]

How middle grade teachers in science and mathematics are prepared should be consistent with the vision of what and how students should learn mathematics and science, in particular the integration of these two fields. In this article a teacher preparation program for middle school mathematics and science teachers that emphasizes the integration of math and science with each other and with technology is outlined. First a theoretical frame-work for the integration of technology is described. Then some examples of uses of technology, such as the use of the Internet, and of interactive and dynamical software that lends itself to establish connections between mathematics and science are given.

The national standards for mathematics (National Council of Teacher of Mathematics, 2000, 1991) and for science (National Research Council, 1996; American Association for the Advancement of Science, 1993) emphasized that educators should prepare students to be literate in mathematics and science, as well as in technology. Yet there is evidence that most middle school classrooms do not use technology appropriately in the teaching and learning process (Jensen & Williams, 1992). One of the goals of the teacher preparation program described in this article is intended to rectify this situation.

---

[21] Flores, A., Knaupp, J. E., Middleton, J. A., and Staley, F. (2002). Integration of technology, science, and mathematics in the middle grades: A teacher preparation program. *Contemporary Issues in Technology and Teacher Education*, [Online Serial] *2*(1). Available: http://www.citejournal.org/vol2/iss1/mathematics/article1.cfm.

TEACHER EDUCATION FOR ARIZONA MATHEMATICS AND SCIENCE (TEAMS)

The mission of TEAMS is to prepare middle school mathematics and science teachers by modeling the use of tools, technologies, and strategies that are consistent with national mathematics and science standards. The standards are the framework for the program and teaching, and also for the teaching, curriculum development, and assessment that prospective teachers are expected to carry out in their own classrooms.

This program supports the reform in middle grades mathematics and science by providing a model for preparing teachers in a way that is consis-tent with the middle school concept. According to this view, the primary fo-cus of middle school is to meet the needs of young adolescents. It is a bridge between the elementary and high school. It is student-centered rather that content centered. Teachers meet the needs of students by incorporating (among other things) flexible block scheduling, interdisciplinary thematic curriculum units, teaming in planning and teaching, and cooperative hetero-geneous grouping. An important assumption of this program is that mathe-matics, science, and technology should be integrated, but both mathematics and science must retain their integrity.

Students in TEAMS are post-baccalaureate individuals who possess de-grees in science, engineering, mathematics, or technology. The TEAMS program is one calendar year and leads to students receiving science or mathematics certification for grades 7-12, a middle school endorsement for grades 5-8, and a master's degree in secondary education.

During the year faculty and students engage in a variety of real-world mathematics, science, and engineering experiences: field trips to various sites within the state; internships in informal education settings such as mu-seums and botanical gardens; and visits to campus research centers. One goal is that prospective teachers observe and use technology while doing science and mathematics in these real-world settings so they can later au-thentically integrate these types of experiences in their future curricula as middle school teachers. Several kinds of technological tools are used during the year: (a) computers, (b) data probes and sensors, (c) multimedia and communication technologies, and (d) graphing calculators.

Fundamental to the process of students becoming teachers in TEAMS is an emphasis on early teenagers' learning with frequent opportunities to work with students in formal and informal settings. Prospective teachers observe and participate in middle school classrooms.

THE USE OF TECHNOLOGY IN LEARNING MATHEMATICS AND SCIENCE

To attain the vision of science and mathematics learning outlined in the Standards, how content is taught is as important as the content itself. In the same way, how the technology is used is crucial if it is really going to help middle school students in their cognitive growth and understanding of mathematical and scientific concepts. According to Pea (1987), cognitive technologies serve two transcendent functions. First, technologies have pur-pose functions. They serve to engage students in the activity of mathemati-cal and scientific inquiry. This provides meaning for engagement, owner-ship of the mathematics and science being learned, and empowerment through the generation of personal agency. Technologies engage students in more powerful scientific and mathematical activity in a way that could not be approached without them. But technologies are not by nature engaging. To achieve this quality, they must be both functional (teachers and students must be able to do with them something that they could not do without them), and they must increase communication and facilitate collaboration.

Second, technologies have process functions. Some of the tools avail-able for students should free up their working memory so that they are able to concentrate on problem formulation and modeling. If a middle school student is bogged down with computing or graphing, the big picture of number systems, functions, families of curves, etc., is lost. Other tools must provide opportunities for exploration and discovery. In a mediated learning environment, some agent (teacher, peer, tool) must bridge the informal knowledge of the student and the formalism of mathematical and scientific structure. Still other tools must provide ways of representing mathematical and scientific models and linking representations to make the underlying commonalties transparent (Lesh, 1979). A single technology rarely has all these process functions. However, a careful selection of tools and software as described in this article can help achieve the necessary complementarity.

Two other features of cognitive technologies are necessary for the de-velopment of coherent mathematical and scientific structures. The first is what Roschelle (1996) called epistemic fidelity. This refers to the require-ment that any teaching tool must reflect and develop understandings that are true to the field of study. Students' mathematical and scientific activity should develop the kinds of understandings that experts in the field would recognize. Two caveats are in order. The road from novice to expert goes through several transformational periods and may not be immediately rec-ognizable as important without an understanding of students' cognitive de-velopment.

Second, the sophisticated knowledge of the expert cannot be handed to students. The path taken is as much a part of expert understand-ing as the final product.

The other necessary feature of cognitive technologies should focus the students' attention on the mathematical structure of the experiences and provide them with a means of communicating their thinking about this structure to others. This is, in its basic form, the engagement of students in mathematical and scientific modeling.

The vision that guides the integration of technology, science, and math-ematics is the engagement of students in activity that elicits the develop-ment of mathematical and scientific models with a coherent epistemological framework. The movement from informal discovery to more formal models marks an authentic transition between the exploratory knowl-edge of the student, and the theoretical knowledge of the expert (Kozu-lin & Presseisen, 1995).

Six principles guided the design, choice of equipment, and software (Middleton & Goepfert, 1996).
1.   Technologies are only tools. Technologies neither supplant the thought processes of students, nor do they make learning fun or easy. Technol-ogies are instruments that should be used judiciously at the proper time in the proper place.
2.   Technologies should enable students do what they could not do without them. When used appropriately, technologies help students expand their zone of proximal development. This can serve to make learning more intentional, powerful, and connected. In addition, computer technologies can represent situations unfeasible with other types of tools.
3.   Technologies must be on hand all the time. The context, social setting, and tools that students use to construct their mathematical and scientific knowledge are inseparable from the knowledge itself. For technologies to be authentically integrated into students' learning activity, they must be available when the question arises.
4.   Tools should facilitate the creation of sharable, modifiable, transport-able models of mathematical and scientific concepts (Lesh & Doerr, 2000). Technologies facilitate the development of public records of thought. These records should be shared as students develop, refine, and test models of mathematical and scientific phenomena. It is crucial that students can modify them, as most models students construct in the beginning are either incomplete, or contain misconceptions. Through discourse, the shared model can be pared down into a workable model that can serve the class as a whole.
5.   Sharing of data/resources should be simple. Technological systems

should be user friendly. The mechanism of communication should not be more complex than the learning process itself.

6.   The setup of the workstations should facilitate collaboration between students. As collaborative tools, technologies are imbedded within the geography, culture, and psychology of the classroom. The setup should facilitate collaborative inquiry, but also engage students in independent exploration.

As can be inferred from these principles, the kind of software and the way it is used are also crucial elements. Common features of the software used in this program are that it can be used by middle grade students; it is user friendly; it is designed for the kind of computers available in schools; and most important, students are in control, telling the computer what to do rather than the computer telling students what to do. The kinds of software used range from general purpose tools to specialized programs for science and mathematics learning. The particular software used can change from year to year. Typically, four or five kinds of technology are used in depth, including computer-based software and graphing calculators. Although pro-spective teachers become quite expert in the use of the technology, the main goal is that their future students use technology to explore concepts and solve problems in science and mathematics. In addition to the examples given in this article, the reader may want to see the examples given by Garofalo, Drier, Harper, Timmerman, and Shockey (2000).

An important emphasis of the integrated approach in TEAMS is that technology is not the only tool to be used. Prospective teachers use it in conjunction with hands-on materials, such as geoboards and polyhedra, and activities such as paper folding. Use of natural objects and outdoor activities are also an important part of integration.

TEAMS AND THE INTERNET

The use of Internet resources is an integral part of the use of technology for prospective middle school teachers. The first tool developed was a web site meant to provide faculty members with a dependable vehicle showcasing their work in TEAMS and other aspects of their professional life. Its function is also to provide access to potential participants and interested colleagues. It serves both for dissemination and recruiting of new candidates for the program.

The program also provides experiences for the participants to learn de-sign and management skills using the web. This aspect of the TEAMS project also serves a double purpose. On one hand, it is a means of disseminating information about our courses and activities. Another function is to let participants learn by actually developing homepages and instructional units, using multimedia applications and authoring tools. The Internet also serves as a tool to facilitate the communication of faculty, mentor teachers, and students in the course of student teaching.

Of course, there are other Internet uses important for middle grades teachers that would be impossible to describe with detail in this article. These include content understanding activities, such as archives, news sources, databases, connections to others, resources for teaching such as video, software, and communications, and electronic portfolio development such as project reports, videos of classrooms, thematic units, internships, and interactive multimedia.

One tool that has been valuable is electronic mail. Students exchange ideas and experiences with their peers and with faculty, both individually and through a listserv. E-mail has provided a forum for them to vent concerns, share experiences, and express feelings and hopes. It also provides a record of teacher growth (Piburn & Middleton, 1998). The interchange of ideas and experiences through the server is especially important during student teaching, due to the fact that students are placed in different schools and could not interact face to face.

EXAMPLES OF OTHER TECHNOLOGIES USED

Interactive mathematics computer programs such as the Geometer's Sketchpad (Jackiw, 1995) and RoboLab (Lego Group, Tufts University, and National Instruments Corporation, 1998) can be used in the middle grades to establish connections between mathematics and science. At the same time, students get acquainted with important aspects of technology. Pro-spective teachers learn to use tools, doing the same kind of exploratory ac-tivities in which their own students in the middle grades could be engaged.

Guided Discovery With Geometer's Sketchpad

An important aspect of mathematical discovery is to learn how to con-jecture and provide convincing evidence. This inductive approach to mathe-matics should be emphasized in the middle grades. A dynamic geometry program such as the Geometer's Sketchpad provides an environment in which prospective teachers can do the same kind of explorations as their own students will do in the future. One example given to TEAMS students is to join the midpoints of consecutive sides of an arbitrary quadrilateral. As teachers change the original quadrilateral they will observe that the in-scribed shape looks always like a parallelogram (Flores, 2001). They can state their conjecture and then provide evidence to convince others about their results.

They can measure angles and opposite sides to verify that, in fact, the inscribed figure shares the same properties as a parallelogram. Teachers can then discuss the analogous process in science of enunciating hypotheses and then gather evidence to confirm or disprove them.

Feedback Systems

An idea central to modern cybernetics and many other fields is that of feedback. RoboLab as two kinds of devices: output devices, such as motors, lights, and sound, and input devices, such as touch sensor, light sensor, and angle sensor. These devices can be controlled with the computer writing procedures in the form of control charts. Prospective teachers use RoboLab to design artifacts with both kinds of devices and write programs that use a feedback loop to control them. Such programs engage students in the fundamentals of robotics, remote sensing, and control.

CONCLUSION

Science and mathematics educators cannot separate the vision of how we should prepare middle grade teachers in science and mathematics from the vision of what and how students should learn science and mathematics in the middle grades. Prospective teachers should have the same kind of experiences integrating science, mathematics, and technology as their future students. One of the goals of the middle school concept is the integration of science and mathematics with other areas. Teachers should experience how technology can be integrated in an authentic way, so that the integrity of both the science and the mathematics is preserved. Different middle schools incorporate to different degrees the ideal of the middle school concept. Prospective teachers can also take part of the approach presented here to implement change and support the necessary reform in mathematics and science teaching over time, regardless of the degree of implementation of the middle school concept in their placement school.

References

  American Association for the Advancement of Science. (1993).
  Benchmarks for science literacy. New York: Oxford University Press.
Flores, A. (2001). *Middle points of quadrilaterals* [Interactive applet]. [Online]. Available: http://www.public.asu.edu/~aaafp/midpointsquadrilateral.html
Garofalo, J., Drier, H., Harper, S., Timmerman, M.A., & Shockey, T.
  (2000). Promoting appropriate uses of technology in mathematics
  teacher preparation. *Contemporary Issues in Technology and Teacher Education,*
  [Online serial] *1*(1), 66-88. Available:
  http://www.citejournal.org/vol1/iss1/currentissues/mathematics/artic
  le1.htm
Jackiw, N. (1995). *The Geometer's Sketchpad* [Computer program]. Berkeley,

CA: Key Curriculum Press.

Jensen, R.J., & Williams, B.S. (1992). Technology: Implications for middle grades mathematics. In D.T. Owens (Ed.), *Research ideas for the classroom: Middle grades mathematics* (pp. 225-243). New York: Macmillan.

Kozulin, A., & Presseisen, B. Z. (1995). Mediated learning experience and psychological tools: Vygotsky's and Feuerstein's perspectives in a study of student learning. *Educational Psychologist, 30*(2), 67-75.

Lego Group, Tufts University, and National Instruments Corporation (1998). *RoboLab*. Lego Dacta.

Lesh, R. (1979). Mathematical learning disabilities. In R. Lesh, D. Mierkiewicz, & M.G. Kantowski (Eds.), *Applied mathematical problem solving* (pp. 111-180). Columbus, OH: ERIC/SMEAC.

Lesh, R. & Doerr, H.M. (2000). Symbolizing, communicating, and mathematizing. In P. Cobb & E. Yackel (Eds.) *Symbolizing and communicating in mathematics classrooms* (p. 361-383). Mahwah, NJ: Erlbaum.

Middleton, J.A., & Goepfert, P. (1996). *Inventive strategies for teaching mathematics: Implementing standards for reform*. Washington, DC: American Psychological Association.

National Council of Teachers of Mathematics. (2000). *Principles and standards for school mathematics*. Reston, VA: Author.

National Council of Teachers of Mathematics. (1991). *Professional standards for teaching mathematics*. Reston, VA: Author.

National Research Council. (1996). *National science education standards*. Washington, DC: National Academy Press.

Pea, R. (1987). Cognitive technologies in mathematics education. In A.H. Schoenfeld (Ed.), *Cognitive science and mathematics* education (p. 89-122). Hillsdale, NJ: Erlbaum.

Piburn, M.D., & Middleton, J.A. (1998). Patterns of faculty and student conversation in listserve and traditional journals in a program for pre-service mathematics and science teachers. *Journal of Research on Computing in Education, 33*(1), 62-77.

Roschelle, J. (1996). Designing for cognitive communication: Epistemic fidelity or mediating collaborative inquiry? In D.L. Day & D.K. Kovacs (Eds.), *Computers, communication & mental models* (p. 13-24). London: Taylor and Francis.

22 HELPING STUDENTS JUMP FROM ARITHMETIC TO ALGEBRAIC NOTATION USING GEOMETRIC REPRESENTATIONS[22]

Three tricks with whole numbers

1) Take the square of a number, add the number, and add the next number. The result is the square of the next number. Example $5 \times 5 + 5 + 6 = 6 \times 6$. Use your calculator to try other numbers. Does it always work?

2) Take any odd number, square it, subtract one. The result is divisible by 8. Example $7^2 - 1 = 8 \times 6$. Try other odd numbers.

3) Take the cube of a number and subtract the number. The result is divisible by 6. Example $4^3 - 4 = 6 \times 10$. Try other numbers to see whether it always works.

Most students are intrigued by tricks like the ones presented and wonder whether they will always work. Some will also wonder why. We can tap into this curiosity to focus on the terms of a numeric relationship, instead of just focusing on computational results. This is an important ability for students in the transition from arithmetic to algebra. This transition can be very difficult for many students. In order to be successful in this transition, students need to

- learn to extract pertinent relations from problem situations and express those relations using algebraic symbols;
- make explicit the procedures they use in solving arithmetic problems;
- consider strings of numbers and operations as mathematical objects, rather than processes to arrive at an answer;

---

[22] Flores, A. and Perkins, I. (2001). Helping students jump from arithmetic to algebraic notation using geometric representations. *Ohio Journal of School Mathematics*, *44*, 23 - 28. Used by permission.

- focus on method or process instead of the answer;
- translate statements about particular numbers to the corresponding generalized statements using variables; and
- write conjectures, predictions, and conclusions (Kieran and Chalouh 1993; Wagner and Kieran 1989).

Geometrical representations of interesting numeric relations can provide a context where teachers can help students develop the abilities needed to make the transition to algebra. Using geometric representations can help students go from statements about particular numbers to the corresponding generalized statements using variables. These representations provide a way to shift students' attention, from the purely procedural approach to numbers, to considering the terms and operations involved in a numerical relationship as entities that are worthy of attention.

In this paper we give some ideas for teachers of middle grades or those teaching the first year of high school on how to help students make the jump from the visual and concrete to the symbolic and abstract. As students write equations representing several cases and compare them, they will see the pattern and will be able to express the equations using variables. Verbalizing the steps and using drawings or other concrete representations can also help students make the connection with the symbols.

Geometric representations can provide fresh meanings for students and ways to see at a glance why the relations between the numbers hold. The drawings and particular numbers used will help students attach meanings to the different terms in the algebraic formula. At the same time, the geometric figures can guide students as they do mathematical thinking of a general nature, that is, that does not just apply to a particular number. For these activities teachers can use inexpensive materials or make their own using grid paper.

Example 1. A puzzle of consecutive squares.

In this activity, cardboard rectangles and squares formed of unit squares are used to help students focus on the relationship between terms in a numerical relation. Rather than just using the numbers to compute an answer, students are guided to observe a pattern in the numeric terms and be able to generalize using variables. Each student receives a set of three pieces. There are at least 5 different sets, so that students can compare results later. Each set consists of a square and two rectangles (see figure 1). Before forming the puzzle, students should describe the relationships among the different pieces. Teachers' questions are important to focus the attention of the students on the different terms and their relation to each other. The teacher needs to give students enough time to think before they answer.

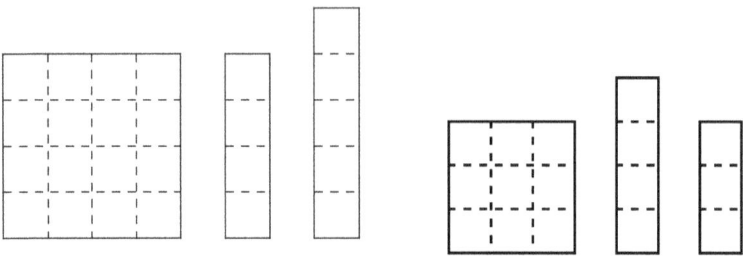

Figure 1a. Three pieces for a puzzle   Figure 1b. Another set

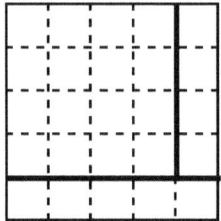

Figure 2. The puzzle formed

Students can label the original pieces 4^2, 4, 5 (or the corresponding numbers), collect everyone's square and make a display like this (Figure 3):

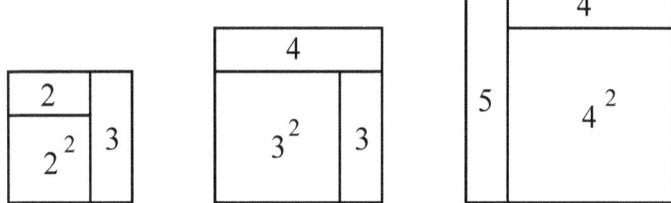

Figure 3. Samples of students' solutions

Students, guided by the teacher, can summarize the results of different puzzles (see Table 1). Table 1 provides some empirical evidence that the general result is true. Organization of empirical evidence is an important step in convincing learners. The next step is to see why the pattern is true for all natural numbers. We can describe each equation on the table as: the square of a whole number plus the number, plus the next number is equal to the square of the next number. If we started out with a square which measured n in both length and width its area would be $n \times n$ or n^2.

Table 1. Consecutive squares

$$2^2 + 2 + 3 = 3^2$$
$$3^2 + 3 + 4 = 4^2$$
$$4^2 + 4 + 5 = 5^2$$
$$5^2 + 5 + 6 = 6^2$$

Students will see that the smaller rectangle has an area of n square units, that its height is n units, and its width is 1 unit. Students should also realize that the bigger rectangle has an area of $n + 1$ square units, a height of $n + 1$ units, and a width of 1 unit. They also realize that the three pieces together can in fact form a square with a side of $n + 1$ units, so that its area is $(n+ 1) \times (n + 1)$ or $(n + 1)^2$. The statement about the sum of the areas of the pieces of the puzzle can now be translated to an algebraic equation $n^2 + n + (n + 1) = (n + 1)^2$. Furthermore, students can then see that the left side is equal to $n^2 + 2n + 1$, the form in which $(n + 1)^2$ is usually expanded.

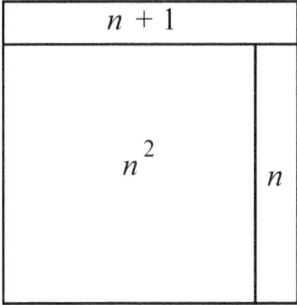

Figure 4. The general case

Triangular numbers

Triangular numbers will be used in the next example, so students need to remember some basic facts. Triangular numbers are sums of the first consecutive whole numbers, for example, the 4th triangular number is $1 + 2 + 3 + 4 = 10$ (see figure 5). Two copies of the same triangular number form a rectangle (see figure 6). The sides of the rectangle are given by consecutive numbers. In general, if we are adding $1 + 2 + ... + n$, the sides of the rectangle would be n by $(n + 1)$. Because the triangular number is only half of the rectangle, we have $1 + 2 + ... + n = n(n + 1)/2$.

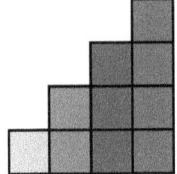

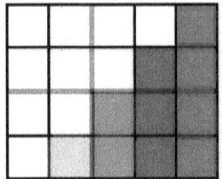

Figure 5. 1 + 2 + 3 + 4   Figure 6. $4 \times 5 = 2 \times (1 + 2 + 3 + 4)$

Example 2. A striking numerical result.

When students take any odd number, square it, and subtract one, the result is always divisible by 8. Students can try different odd numbers in a systematic way, and organize data as in table 2. Some students may recognize the triangular numbers in the last column of table 2. Based on this observation they may predict what the next result will be.

Table 2. Squares of odd numbers, minus one

Odd number	squared	minus one	divide by eight
1	1	0	0
3	9	8	1
5	25	24	3
7	49	48	6
9	81	80	10

The following activity is geared to help students understand why when we subtract one from the square of an odd number the result is always divisible by eight. Students are given a drawing of a square (with a side of odd length) that is missing a square (figure 7). A variety of examples is important. Asking students to list their observations is informative and many times their observations are surprising.

Directions and questions for students: Copy the square minus one unit onto grid paper and then cut off the shaded section. Can you make a rectangle using those two pieces? (Figure 8). What is the length of this rectangle? What is the width of this rectangle? How many more units is the length compared to the width? Why are both the length and the width even numbers?

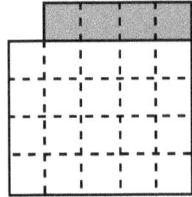

Figure 7. $5^2 - 1$

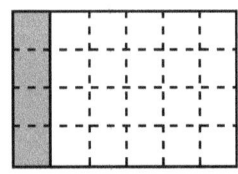

Figure 8. $5^2 - 1 = 4 \times 6$

In this case, they will see that $5^2 - 1 = 4 \times 6$. Students can use grid paper to cut out another rectangle using grid paper that has the same width and length as this rectangle. Students divide the new rectangle into 4 smaller rectangles with equal area (see figure 9). Students should notice that the small rectangles have the property that one side is one unit longer than the other. This is the kind of rectangle formed by two copies of a triangular number (see figure 10). Thus, the rectangle is equal to eight triangular numbers. Therefore, the square of the odd number minus one is also equal to eight triangular numbers. Therefore, it is divisible by eight.

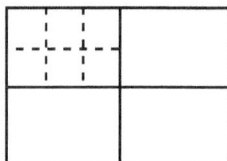

Figure 9. $4 \times (2 \times 3)$

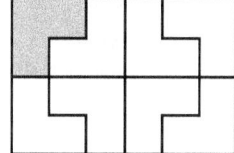

Figure 10. Eight triangular numbers

By doing activities as described above, students will have meanings they can attach to the symbols in an algebraic demonstration. An odd number can be expressed as $2n + 1$, where n is an integer. Squaring it we have $(2n + 1)2 = (2n + 1) \times (2n + 1) = 4n^2 + 4n + 1$, subtracting one we have $4n^2 + 4n$. This can be written as $4 \times n \times (n + 1)$ [four rectangles]. It also can be written as $8 \times n \times (n+1)/2$ [eight triangular numbers]. We can summarize this example by the formula $(2n + 1)^2 - 1 = 8 \times n \times (n + 1) / 2$.

Example 3. A cube with a column removed.

We can also use cubes to understand other interesting number relationships. We saw that if you take any number and raise it to the third power and subtract the original number, the result is divisible by six. Let us use 4 as our starting number. Figure 11a represents a 4 by 4 by 4 cube with a column deleted along an edge, that is, the cube minus the number. We can build a rectangular block with the remaining pieces by moving one slice and put it on top (figure 11b). The dimensions of the block are now 3 by 4 by 5. Observe that they are consecutive numbers. When we have three consecutive numbers, at least one of them has to be even, and one has to be divisible by three. So the product of the consecutive numbers is divisible by 2 and by 3, and therefore is divisible by 6. In general, if you take from a cubic number $n3$ a column of n unit cubes, the remaining volume will be $n3 - n$. If you rearrange the cubes by moving a slice in the same way, you can form a block with dimensions $n - 1$ by n by $n + 1$, that is, three consecutive numbers. Therefore $n3 - n$ is divisible by 6. This example can be used to show students how by rewriting algebraic expressions, as in this case factoring $n^3 - n$, we can obtain additional insights into the relations of the quantities involved.

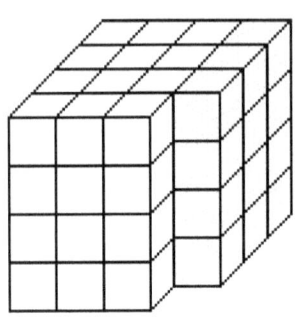

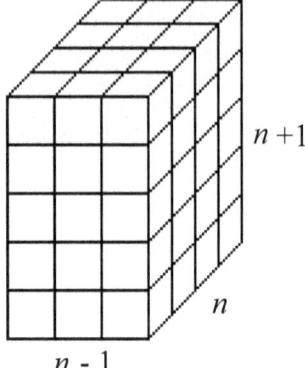

Figure 11a. $n^3 - n$ Figure 11b. $(n - 1) \cdot n \cdot (n + 1)$

Conclusion.

All too often we require students to make the transition from arithmetic to algebra too quickly. Let us give students the time and opportunity to make the transition from the concrete and particular to the abstract and general. Concrete representations of numbers and their relations in the form of manipulative materials, puzzles, and visual displays can help students give meaning to the expressions using symbols. In addition, students can explore relations that can be generalized. By using and extending the patterns they observe among numbers, they will find it easier to use variables to express the relation in general.

In algebra and beyond, students need the ability to deal with symbols for variables rather than just symbols for numbers. Algebraic notation will help them reason about statements that apply to all numbers or to numbers in a specified set, rather than about statements that apply to particular numbers. Algebraic notation has the power to carry much of the weight of thinking when deriving or proving mathematical results. Its abstractness makes it a powerful tool across contexts.

However, in the beginning, while students develop their skills with the symbols, it is important that they connect meaning to what the manipulation of symbols represent, so that their manipulations do not become senseless. Geometrical representations can provide guidance and understanding as to why each step or term in a chain of algebraic manipulations or expressions is correct. Later, when students develop their skill, and their algebraic (or symbol) sense, geometric representations can be additional sources of discoveries and inspiration.

References

Kieran, Carolyn, and Louise Chalouh. Prealgebra: The transition from arithmetic to algebra. In *Research ideas for the classroom: Middle grades mathematics*, edited by Douglas T. Owens (p. 179 - 198). Reston, VA: National Council of Teachers of Mathematics, 1993.

Wagner, Sigrid, and Carolyn Kieran. *Research issues in the learning and teaching of algebra*. Reston, VA: National Council of Teachers of Mathematics, 1989.

23 INCLINED PLANES AND MOTION DETECTORS: A STUDY OF ACCELERATION[23]

Summary
Students work in cooperative groups, one team per ramp. They roll balls down inclined planes, collect data with the help of an electronic motion detector, and represent the data with a graphing calculator, to explore such concepts as mass, gravity, velocity, and acceleration.

Grade level: 7 - 12

Mathematics Skills/Concepts
Measure of distance, angle and height of ramp; computation of velocity and acceleration; slope of the distance over time depicts velocity; slope of velocity over time depicts acceleration; sum of odd numbers is a square.

Science Concepts/Processes
Velocity, acceleration, gravity, relation between height of release and terminal velocity, mass, friction.

Prerequisite skills
Students need to be familiar with graphing calculators and how to read and trace graphs on them. This activity is not meant to be an introduction to the use of motion detectors and graphing calculators to study motion; students should have prior experiences with them. This activity provides an

---

[23] Flores, A. and Turner, E. (2001). Inclined planes and motion detectors: A study of acceleration. *School Science and Mathematics, 101*(3), 154-161. Reprinted by permission of Wiley and School Science and Mathematics Association.

opportunity to expand upon and deepen their understandings of motion.

Objectives

Process Objectives. The students will work cooperatively, delegate and share responsibility; take care of equipment, store and use equipment appropriately; record data in an organized manner; use measurement devices (rules, timers, Calculator Based Laboratory (CBL), scales) appropriately to arrive at accurate measurements; discuss experiments and questions among themselves; communicate observations and experiences to the rest of the class.

Content Objectives. Observe the relationships between mass, height of ramp, and point of release based on their experiences; determine what elements will affect acceleration; record data points from their experiments; use distance and time data to compute velocity; use velocity data over time to compute acceleration; create graphs to represent their data; explain and interpret the graph for the class. Explain the meaning of the slope, the meaning of the various data points; justify their conclusions using their data; use the concepts of distance, time, velocity and acceleration appropriately; determine factors which may have impacted the accuracy of their experiment (friction, error in measurements, etc.).

Rationale

Content background

One of the most popular stories about Galileo (1564 - 1642), is that he threw objects from the Leaning Tower of Pisa to demonstrate that objects of the same kind, but with different masses reach the ground at essentially the same time. Although there are no contemporary records that Galileo did this in a public event, and the first account appeared 15 years after Galileo's death and more than 60 years after the time when Galileo taught at Pisa, the experiment is quite plausible. Galileo states in his book *Two New Sciences* that a hundred pound iron ball and one of just one pound falling from the height of a hundred braccia arrive at the same time, and adds "You find, on making the experiment, that the larger anticipates the smaller by two inches; that is, when the larger one strikes the ground, the other is two inches behind" (p. 68). Such an experiment would constitute a demonstration, but would not provide an adequate setting to study accelerated motion. To study accelerated motion, Galileo used inclined planes to slow down free fall, and minimize the effects of air resistance. At that time, measuring instruments were not very sophisticated. To measure time Galileo used "a large pail filled with water and fastened from above, which had a slender tube affixed to its bottom, through which a narrow thread of water ran" (p. 170). In spite of these limitations, Galileo was able to corroborate experimentally his conclusion, obtained through mathematical reasoning, that "if a moveable descends from rest in uniformly accelerated motion, the spaces run through in any times whatever are to each other ... as the squares of those times" (Proposition II. Theorem II, p. 166).

In the past, due to the difficulties of measuring time and position at the same time, teachers used ingenious devices to demonstrate this result. As Galileo noticed, the differences between successive squares of whole numbers are given by the odd numbers 1, 3, 5, 7, ... An object with uniformly accelerated motion will traverse in equal times segments proportional to the odd numbers. Therefore, if we place objects along the path of a heavy rolling ball, with distances between them in the ratio of the odd numbers, so that when the ball touches them slightly they make a sound, we will hear sounds rhythmically produced in equal intervals of time. Other devices used by teachers include rattles or bells attached to a string at distances that are proportional to the odd numbers. The string is held vertically and dropped; the sounds of the bells hitting the ground are produced also in equal intervals of time. The second author participated as a high school student in an even more spectacular demonstration involving a long rope with water balloons attached to the rope at intervals proportional to the odd numbers. The left part of figure 1 illustrates how the balloons are attached to the rope. The rope is held from above. The distances between successive balloons are indicated. The rope was held vertically from the second floor and dropped. The splash sounds of the water balloons as they hit the ground were produced in equal intervals of time. In contrast, if balloons are equally spaced on the rope (right part of figure 1), the sounds succeed each other at shorter and shorter intervals of time.

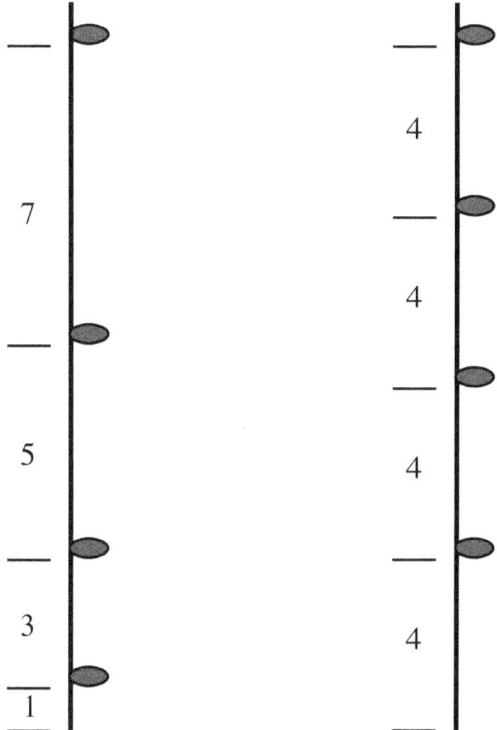

Figure 1. Water balloons hanging from a rope.

Technology Background

These activities can be conducted using a motion detector that works in conjunction with a Calculator Based Laboratory and a graphing calculator. A program is used to control the data collection, and represent the information on the calculator. (We used TI-83, CBL, and Vernier's motion detector, but the same activities can be done with other calculators and devices). Information on how to use these devices is readily available in books and in documents that can be downloaded from the internet. A good introduction is the *CBL System Guidebook* available for free from the internet (Texas Instruments Instructional Communications, 1997).

There is no need for the teacher to write the program or even to enter it by hand into the calculator. Similar programs can be downloaded for free from the internet (see for example http://www.ti.com/calc/docs/arch.htm) or are available on disk with some of the books about CBLs. The programs can then be transferred into the calculator from the computer with the help of TI-Graph Link (1998) software, also available for free download from the internet). The programs then can be shared from calculator to calculator using the communication port at the lower part and the Link key. The CBL and the calculator are linked using the same cable (see figure 1). Probably the biggest source of problems (and frustration) when using the CBL and graphing calculator is the link between the two. Press the cable firmly into the port at the bottom of the calculator.

Figure 1

These activities assume that students are already acquainted with the use of CBLs, motion detectors, and graphing calculators to study moving objects. For example, an activity like *Take a hike* in <u>Exploring mathematics with the CBL</u> (1997), or like the ones described by Doerr, Rieff, and Tabor (1999) would provide an adequate preparation for this one. We give here only a brief description of the use of these devices.

<u>The program Hiker for TI-83</u> (short version). This program is appropriate to collect data about objects that do not move too quickly. It displays the data on the calculator screen (almost) in real time. The program will start collecting data once you press the enter key. This shortened version skips the presentation screens and does not check the link connection. The program HIKER will instruct the CBL to collect 60 data, every 0.1 seconds.

Once you have stored the program into your calculator, follow these steps to use it:

1. Press the PRGM key on your graphing calculator.

2. Use the down arrow key to scroll down the list of programs until you highlight program HIKER (figure 2a).

3. Press ENTER to select program HIKER. The calculator will show the display on figure 2b. The computer will then begin the program (figure 2c).

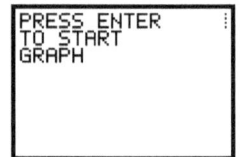

Figure 2a   Figure 2b   Figure 2c

Place the ball on the ramp about 50 cm from the motion detector (see figure 3). Follow the steps above. When the calculator tells you to press ENTER, do so and then release the ball. The motion detector should start buzzing, indicating that it is collecting data. The graphing calculator will display the corresponding graph of positions of the ball at given times. (The motion detectors we are using have a minimum range of about 30 cm and a maximum accurate range of about 10 m.)

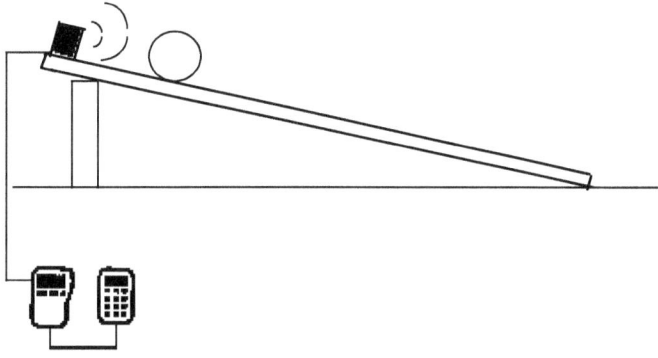

Figure 3

Research/Professional Recommendation Background

The *National Science Education Standards* (1996) recommend for the grades 5 - 8 the study of motion. "The study of motions and the forces causing motion provide concrete experiences on which a more comprehensive understanding of force can be based in grades 9 - 12" (p. 149). The National Council of Teachers of Mathematics (2000) emphasizes the importance of integrating the use of up to date technology in the teaching of mathematics. Technology should be used "with the goal of enriching students' learning of mathematics" (p. 25). Technology is also crucial for learning science concepts. It is important for students to understand the reciprocal nature of science and technology. On one hand, science helps drive technology. On the other hand, "technology is essential to science, because it provides instruments and techniques that enable observations objects and phenomena that are otherwise unobservable due to factors such as quantity, distance, location, size, and speed" (National Science Education Standards, p. 166). The instruments used in these activities are an example of how technology also provides tools for investigations, inquiry, and analysis. The activities are also an example of how technology enables students study the objects that move rapidly (American Association for the Advancement of Science, 1993).

Instructional Model

The students will be organized in several small teams of 3 to 4 students, one group per ramp. Students within each team will be assigned roles for the activity, such as materials manager, data recorder, experimenters, materials builder, group speaker. The lesson proceeds in a cycle, where students first interact with the balls rolling down the ramp in an informal exploration. Then they share their observations with the rest of the class. This sets the stage for a discussion where concepts are developed and a more focused investigation is conducted. A final debriefing and discussion will permit children to reflect on and interpret their discoveries.

Lesson Outline

Time needed: This experiment will require at least two 45-minute class periods.

Materials/Advance Preparation: Low friction inclined planes, rulers, balance scales, masses, balls of different masses (steel, wooden, plastic, and rubber balls), Calculator Based Laboratory (CBL), graphing calculator, motion sensor, student data sheets. Let students build the inclined plane with the motion detector on top as shown in figure 3 (they may have to tape the motion detector to the inclined plane).

Initial Student Exploration

To begin this investigation the teacher will present the materials available, and explain the methods students can use to collect and record data (i.e. CBL, calculator, timing the length of the balls trip down the ramp, and measuring the distance the ball travels, the height of the ramp, and the mass of the ball). To engage students in the experiments, the teacher asks them to *predict* what will happen. The teacher can ask questions such as: If we keep the slope of the ramp constant, what balls will roll down faster? Why? What happens if we increase the slope of the ramp? Will the balls reach the same speed if we release them from the middle of the ramp? Students share their predictions. Usually students are eager to test their predictions. To guide the data collection, the teacher poses the following questions to the students, explaining that each group will focus their explorations around each of the questions. If time is limited, the teacher can assign each team the responsibility to answer one of the three questions:

- What do you notice about the motion of balls of different masses as they roll down a ramp?
- What do you notice about the motion of balls if you release them at different points on the ramp?
- What do you notice about the motion of the balls as you change the slope of the ramp?

Student Data Presentation

Each team will be given 5-10 minutes to present their observations to the rest of the class.

Teacher Led Discussion - Concept Invention

Teacher will lead the discussion by asking about concepts that have arisen from students observations. Teacher's questions can focus their attention on concepts such as velocity, distance traveled, acceleration, and mass. The teacher can also vocabulary to make more precise the discussion. The teacher may ask about ways to compute velocity over time, or what is the effect of mass on velocity and acceleration. Some of the concepts involved in this activity will be difficult for students at first, and so it is very important to give examples of the concepts in terms that the students can understand and connect with when they travel in cars, ride their bicycles, or observe air planes. For example when riding a bike downhill, compare what happens if the slope is very steep or if it is just a gentle slope. Why do planes dive when the pilot wants to attain a big speed?

The teacher will help the students formulate two questions to explore further in the next experiment. These questions may come up naturally based on the students' experiences and observations during the exploration activity. If these questions do not arise, the teacher will pose the questions to the students. Both questions will focus on the concept of acceleration, and what seems to influence acceleration. As the teacher forms each question, he/she will state it in terms of the experiment: Do you think that a ball with more mass, like this one, would accelerate faster as it travels down the ramp, or would it have the same rate of acceleration as a ball with less mass? Do you think that if we keep raising one end of the ramp, and therefore increasing its slope, the ball will accelerate more?

Students will focus their attention on the following questions:

• What is the effect of the mass of the object on the acceleration of the object? Why do you think so?

Often students answers at this point still reflect common misconceptions, such as that the bigger steel ball will accelerate much more than the smaller steel ball.

• What is the effect of the height and slope of the ramp on the acceleration of the object? Why do you think so?

Generally students predict that when the slope is increased, the velocity will increase also.

The teacher will discuss with the students how they could possibly investigate these two questions. The teacher will help them to design an experiment to test each question. The two experiments are: a) use the same ramp kept at a constant height, and then roll balls of different masses down the ramp; b) use one ball, keeping the mass constant, and change the height (and therefore the slope) of the ramp, and roll the ball down ramps of different heights (and slopes).

Focused Experiment

Students will conduct this controlled experiment designed by the teacher and students to further explore the relationships among mass, height of ball release, and acceleration. Students will complete this task in teams of four. The teacher will assign half of the teams the first experiment and the other half of the teams the second experiment.

All teams must use the CBL data collecting device to collect data for their experiment. They should use the motion detector to track the motion from each trial. From each trial, the students should make a quick sketch of the appearance of the graph, and then record 8 sets of data points of their student recording sheet. The sets of data points will include a time, and the corresponding distance at that time. Students will record their data points based on pre-established time intervals (in our case at every 0.1 second). The teacher may wish to have all students record data in some experiments, and just assign one student to record data on other occasions.

Table 1 Distances, Velocities, Acceleration

Time	Distance	Velocity	Acceleration
0	0		
		1	
1	1		2
		3	
2	4		2
		5	
3	9		2
		7	
4	16		2
		9	
5	25		2
		11	
6	36		2
		13	
7	49		2
		15	
8	64		

Velocity and acceleration from a distance table. Once students have collected all data, they will work in teams to calculate velocity and acceleration based on the distance and time data. The teacher may illustrate with an example how information about distances taken at constant time intervals can give us information about velocity. Table 1 illustrates the case of distance increasing as the square of time. To obtain the distance traveled in each time interval, simple take the difference between distances. To obtain the average velocity divide this distance by the length time interval (in this case 1). Once we have the average velocities for each interval, we can compute the acceleration, that is, the rate of change of the velocity. To compute the average acceleration from one interval to the next, take the difference between velocities, and divide by the time difference (in this case 1).

In a similar way, for each of their trials, students should extract eight sets of time-distance data points (at equal time intervals), which will allow them to compute seven velocity points (let students figure out how many). From those seven velocity points, they can then compute six acceleration points (let students figure out the number).

After students have computed velocities and accelerations for each of their trials, they should create a graphical representation of their data. Students may choose to create these graphs in different formats, which is acceptable, as long as we are clear that we cannot then compare one graph to another directly. If we wish to compare the graphs, the teacher may want to set more limitations and standards about the size, the type, and the intervals on the graph.

Students will then use their data and their graphs to form a preliminary conclusion about their experiment. Looking back at the original question, what can they say? What have they learned about how mass/height seems to influence acceleration? What else have they learned? Students will record these conclusions, and their thoughts about what they discovered during the experiment in writing. Although group member will discuss their conclusions and what they have learned as a group, each student will complete an individual written reflection.

Whole Group Discussion - Debriefing

After students have completed the group experiment and the individual reflection, the teacher will engage the students in a discussion / debriefing of the experience. The teacher will first ask all teams to share the results of the experiments. Students will present their graphs and describe the results of their experiment to the class. They will also share what they learned about what seems to impact acceleration, and what else they learned about forces and motion during this experiment. Other students in the class will be encouraged to comment upon and ask questions about the experiments each team conducted.

As students present, the teacher will pose questions to the class to help them compare one experiment to another, and to help them think more critically about the scientific concepts. Questions may include the following (we write in parentheses some answers students tend to give):

- Two groups did the same experiment varying the slope of the ramp, what do you notice about their results? (The steeper one went faster.)
- These two teams both experimented with masses of different sizes, but made of the same material. How did the mass affect the velocities and/or accelerations? (If the material is the same, there was not so much of a difference; I had expected the heavier one to go much faster than the other.)
- How can you explain why this team obtained a different result from other groups? (Because they let the ball off at a different point of the ramp.)
- Do you agree with the conclusion that this group formed, based on the data from their experiment? (Yes, the velocities are bigger for the steeper ramp.)
- What similarities do you notice between the data from these groups? (In both cases the velocities seem to grow at a steady rate.)
- Looking at this data, what seems to have the largest impact on acceleration? (The steepness of the ramp.)
- What factor does not seem to have such a large impact? (The mass of the ball, as long they are of the same material. If materials are different, then friction becomes a bigger factor.)
- What conclusion would you draw from these data? (Mass does not have such a big role as I thought in the beginning.)
- What other experiments could you conduct to help you learn more? (Maybe keep the height constant, and use ramps of different sizes to change the slope.)

As students discuss these and other questions that arise, the teacher will again stress the scientific and mathematical concepts, helping the students to clarify their understandings of these concepts by explaining them in light of the experiments the students just completed.

Evaluation
The teacher will collect the graphs, student sheets, and student written reflection for assessment purposes. As the teacher reviews the students work, he/she will focus on whether the students were able to
- record what they did in their experiments;
- carry out a logical, valid experiment (controlling for variables);
- record sets of data points;
- calculate velocities and acceleration based on their data points;
- represent their data in a graphical form; and
- draw logical conclusions based on their data.

The teacher will also use observation notes that he/she made as students were working and presenting their data for assessment. As the teacher observes students working, he/she will focus on the following: How do team members share responsibilities? Do team members carry out their assigned role? Do all team members participate? How do the teams seem to solve problems that arise?

Conclusion

Galileo's genius allowed him to see that movement along inclined planes would give him a better understanding of free fall and accelerated motion. However, he also was a skilled experimenter and was able to use smooth balls of metal, and long inclined planes that were almost frictionless. Those skills and equipment would not be available to students in the middle grades or high school. However, nowadays, relatively inexpensive technology such as graphing calculators, Calculator Based Laboratories, and ultrasonic motion detectors allow our students to measure time and position with ease and graph the movement of the balls as they roll down inclined planes. The new technology provides access to students at a younger age to experiments and data from which they can study accelerated motion and develop the corresponding concepts.

References

American Association for the Advancement of Science. Benchmarks for science literacy. New York, NY: Oxford University Press, 1993.

Brueningsen, C., Brueningsen, E., & Bower, B. Math and science in motion: Activities for middle school. Dallas, TX: Texas Instruments. Available http://www.ti.com/calc/docs/msmotion.htm

CBL Explorations in Algebra for the TI-82 and TI-83. Erie, PA: Meridian Creative Group, a Division of Larson Texts, 1996.

Carlson, Ronald J. & Winter, Mary J. Transforming functions to fit data. Berkeley, CA: Key Curriculum Press, 1998.

Doerr, H. M., Rieff, C., & Tabor, J. (1999). Putting math in motion with calculator-based labs. Mathematics teaching in the middle school, 4(6), 364 - 367.

Exploring mathematics with the CBL. [On line]. Texas Instruments, 1997. Available http://www.ti.com/calc/docs/act/pdf/cblworkm.pdf

Galilei, Galileo. Two new sciences (translated by Stillman Drake). Madison, WI: University of Wisconsin Press, 1974.

Lund, Charles, Anderson, Edwin, and Zacny, Carol (Eds.) Graphing Calculator Activities: Exploring Topics in Algebra 1 and II. Dale Seymour Publications, 1997.

National Council of Teachers of Mathematics. Principles and standards for school mathematics. Reston, VA: National Council of Teachers of Mathematics, 2000.

National Science Education Standards. Washington, DC: National Academy Press, 1996.

Randall, Jack. Sensor Sensibility: Algebra explorations with a CBL, TI-82 or

TI-83, and Sensors. Berkeley, CA: Key Curriculum Press, 1997.
Texas Instruments Instructional Communications. CBL System Guidebook. Texas Instruments, 1997. [On line] Available http://www.ti.com/calc/pdf/gb/cblguide.pdf
TI-Graph Link (computer software). Texas Instruments, 1998. Available http://www.ti.com/calc/docs/link.htm

Equipment
CBL System Calculator Based Laboratory. Dallas, TX: Texas Instruments.
Graphing calculator TI-83. Dallas, TX, Texas Instruments.
Ultrasonic Motion Detector. Portland OR: Vernier Software.

24 SURFACE AREA OF THE SPHERE: A HEURISTIC ARGUMENT[24]

ABSTRACT: The surface area of the sphere is obtained by using truncated cones that have the same lateral area as the lateral area of the corresponding rings of the cylinder that is tangent to the sphere, so that the sum of the areas of the cones is constant. As the cone sections become smaller and smaller, they approximate better and better the sphere. Therefore, the lateral area of the tangent cylinder is equal to the area of the sphere.
KEYWORDS: Surface area, heuristic arguments, limiting process, integral calculus.

INTRODUCTION

In this paper we use sections of tangent cones to the sphere to approximate its surface area. This approach can enrich the learning for students in calculus courses in two ways. On one hand it provides an alternative approach that will help students understand why the formula for the surface area of the sphere works, and how it can be obtained. On the other hand it also gives them the opportunity to deal with a limiting process that is slightly different from the other limiting processes used in calculus.

Approximating the surface of a sphere by truncated cones to find its area is of course not new. Archimedes [1] used inscribed and escribed truncated cones to show that the area of the sphere is equal to four times the area of its greatest circle. One such circle is for example the circle encompassed by the equator. If r is the radius of the sphere, Archimedes result can be given

---

[24] Flores, A. (2000). Surface area of the sphere: A heuristic argument. *PRIMUS*, *10*(4). 345-350. Reprinted by permission of Taylor & Francis (http://www.tandfonline.com).

in modern notation as $A = 4\pi r^2$. This value is equal to the lateral area of the cylinder (that is, without top and bottom) that is tangent to the sphere and has the same height as the diameter of the sphere (Figure 1). To see that 4π $r2$ is the lateral area of the cylinder we cut it and lay it out to obtain a rectangle with dimensions $2\pi r \times 2r$ (Figure 2). That the area of the sphere is the same as the lateral area of the cylinder is a result that students find surprising. In this article we will use truncated cones that are slightly different than the ones used by Archimedes to help students see *why* these two areas are equal.

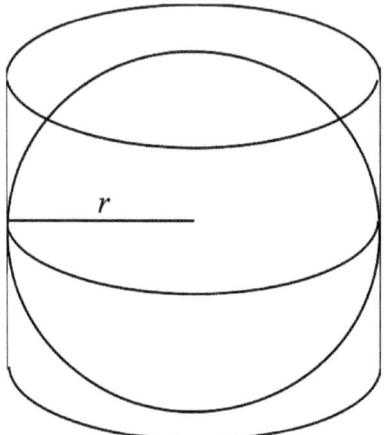

Figure 1. The cylinder tangent to the sphere and height equal to 2r.

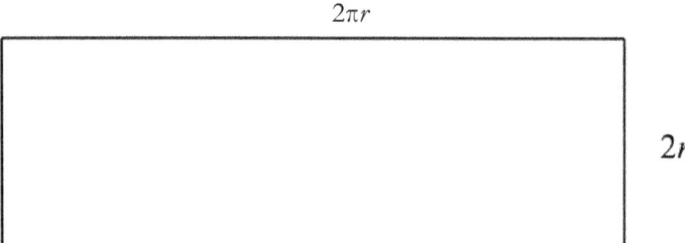

Figure 2. The lateral surface of the cylinder laid out.

A TRUNCATED CONE WITH THE SAME AREA AS THE CYLINDER

We will consider first the case of the truncated cone that is tangent to the sphere at the 30° parallel, and lies between two parallel planes, one through the equator and another through the pole (see Figure 3). This is what Pólya [3] has called a *leading particular case*, because it will lead the way to the general solution. The generator AB of the cone (see Figure 4) is divided in two equal parts by the circle of tangency (notice, however, that the arch of the sphere section is not divided into equal segments). The radius of the circle of tangency is equal to $r\cos(30°)$; the length of the generator is $r/\cos(30°)$. That is, the product of length of the generator times the radius of the circle of tangency is equal to the product of the height of the cylinder times its radius. The truncated cone has a surface area given by length of the generator × circumference of circle of tangency, that is, $r/\cos(30°) \times 2\pi \times r \times \cos(30°) = 2\pi\, r^2$. The lateral area of the cylinder tangent to the sphere with height r is also $2\pi^2$.

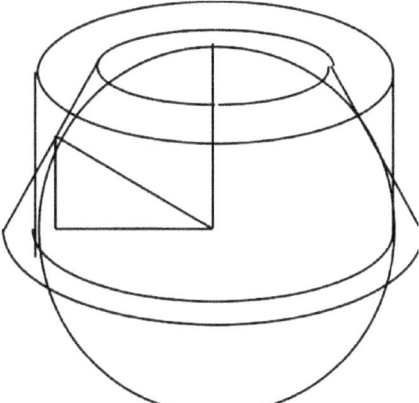

Figure 3. Truncated cone tangent to the sphere at the 30° and tangent cylinder.

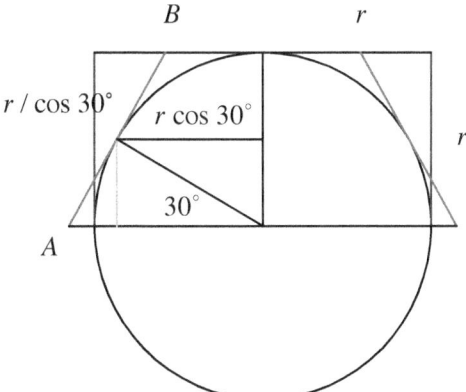

Figure 4. Cross section of cone tangent to sphere at 30° parallel.

APPROXIMATING THE SPHERE BY CONES

We will use the same idea as in the previous case to find thin truncated cones that approximate the surface of the sphere and have the same area as the lateral area of the corresponding slice of the cylinder that is tangent to the sphere. That is, we will find truncated cones tangent to the sphere such that the product of length of the generator times the radius of the circle of tangency is equal to the product of the height of the corresponding slice of the cylinder times its radius. First, the surface of a sphere and its tangent cylinder are cut by planes parallel to the equator. The sphere strips so formed are approximated by truncated cones, that are tangent to the sphere at their circular section of mean radius, that is, half way between the bottom and the upper rim. Figure 5 shows one of such cones between two planes with a distance z between them. The mid circle of the truncated cone is tangent to the sphere at a parallel of latitude $ø$. The cones are not necessarily joined exactly together, and the fit of the cone that "covers" the section of the sphere around the pole is not very good when there are only a few subdivisions, but this will not cause any problem.

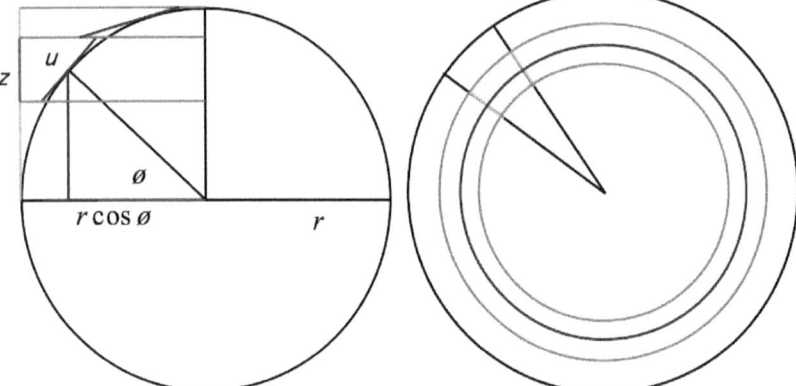

Figure 5a. Lateral cross section   Figure 5b. View from above

We will show that the area of the truncated cone is the same as the lateral area of the strip on the cylinder tangent to the sphere obtained by axial projection (same property as the special case of the truncated cone tangent at the sphere at the 30° parallel line, whose area was equal to that of the whole cylinder). Figure 6 shows a section of the cone and the corresponding section of the cylinder. Let x be the length of the arch of tangency of the cone to the sphere. The area of the conical trapezoid is $x \times u$, and the area of the rectangle on the cylinder is $y \times z$. We have that $y : x = 1 : \cos ø$, and on the other hand $u \times \cos ø = z$. Therefore the conical trapezoid and the cylinder rectangle have the same area (see Melzak [2] for a slightly different heuristic argument).

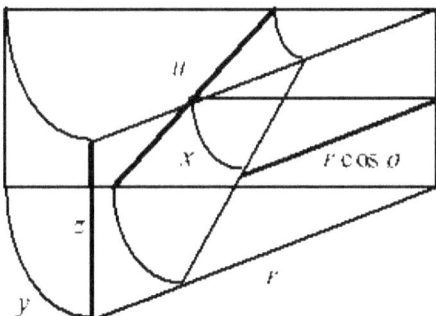

Figure 6. Conical trapezoid and cylinder rectangle.

Thus the area of the truncated cone is equal to the lateral area of the corresponding strip on the cylinder tangent to the sphere. The sum of the areas of the truncated cones is therefore equal to the lateral area of the cylinder tangent to the sphere, which is $2\pi r \times 2r = 4\pi r^2$. By increasing the number of parallel planes and making the distance between them smaller and smaller, the truncated cones will approximate better the surface of the sphere. Therefore the area of the sphere is equal to $4\pi r^2$.

REFERENCES

1. Archimedes. 1953. *The works of Archimedes*. Edited by Thomas L. Heath. New York: Dover Publications.
2. Melzak, Z. A. 1973. *Companion to concrete mathematics*. New York: Wiley.
3. Pólya, George. 1962. Mathematical discovery: On understanding, learning, and teaching problem solving, Vol. 1. New York: Wiley.

25 THE PARABOLA AS THE ENVELOPE OF A FAMILY OF STRAIGHT LINES[25]

ABSTRACT: Curves are obtained by stitching or drawing straight lines that join points on two lines that satisfy a certain relationship. A graphing calculator and a dynamical geometry program are also used to obtain families of lines with the same characteristics. Then we prove that a parabola is indeed the envelope of the family of straight lines.
KEYWORDS: Parabola, envelope, curve stitching, dynamical geometry.

INTRODUCTION

Here we give an example of how students can use calculus ideas to provide a mathematical explanation of why the parabola is the envelope of a family of straight lines of the kind commonly used in string designs. An envelope of a family of lines or curves is a curve that is tangent to all of them. We show how concrete and familiar representations of curves can be connected to a more abstract and symbolic approach to these curves.

Previous experiences with curves in the context of string or lines designs can be the basis to gain familiarity with curves of mathematical nature (Somervell [9]). However, many times students who have encountered curves in such activities may not have had the opportunity, nor the mathematical tools, to understand why the curve formed by the straight lines is indeed a parabola (or other kind of curves, depending on how the straight lines are formed). On the other hand, all too often, students in college courses miss completely concrete approaches to curves, and their first experience with curves such as

---

[25] Flores, A. (2000). The parabola as the envelope of a family of straight lines. *PRIMUS, 10*(3), 257-266. Reprinted by permission of Taylor & Francis (http://www.tandfonline.com).

the parabola are when they see them represented in a highly symbolic form, such as $y = x^2$, and the corresponding graph.

In the first part we describe concrete activities that give rise to curves as envelopes, including use of dynamical geometry programs. These concrete experiences are within the reach of younger students also, however, we have found that university students benefit from such activities too. Then we use a basic property of tangents to parabolas to see why the family of straight lines forms a parabola, and finally we see how to obtain the equation of the curve that is tangent to a specific family of straight lines. Although the method is used here only for the parabola, it can be used for other families of straight lines or circles that form other curves as envelopes (Boltianskii [1]).

CURVES FROM STRAIGHT LINES.

String designs can form interesting curves. Students can create beautiful geometrical patterns by stitching or by drawing straight lines on paper. The easiest envelopes to stitch or draw are probably parabolas (see Figure 1). We will describe how to draw them on paper; readers interested in stitching them can find instructions in Somervell [9] or Millington [8]. Figure 2 shows some examples where arcs of parabolas are formed, and are combined to form other shapes.

Figure 1. Parabolas out of straight lines.

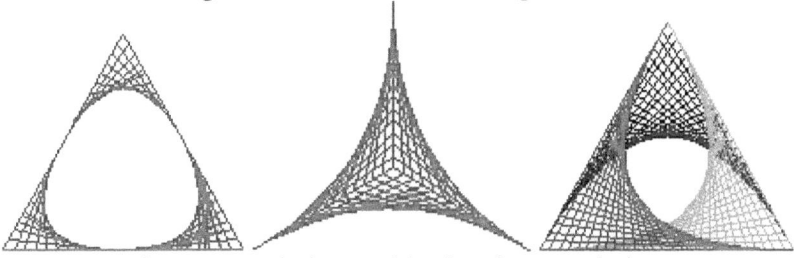

Figure 2. Parabolas combined to form art designs

The parabola is formed by joining corresponding equally spaced points along two line segments. One way to do it is by joining numbers that add to 10 (see Figure 3). The resulting pattern will suggest a curve.

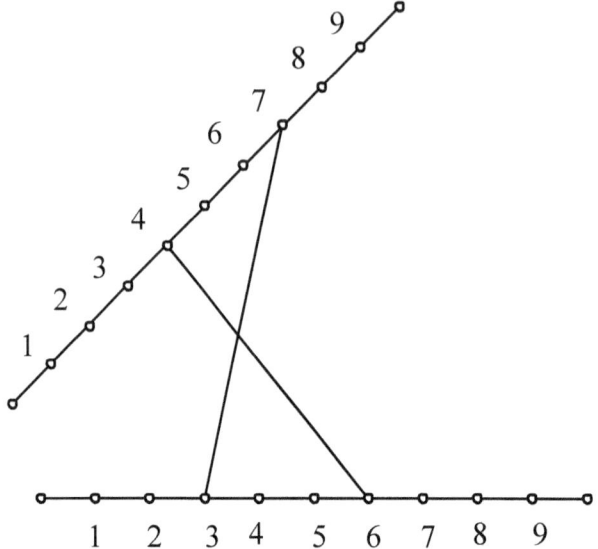

Figure 3. A procedure based on numbers.

Dynamical geometry programs, such as The Geometer's Sketchpad (Jackiw [6]) and its web component JavaSketch (Jackiw [7]), are great tools to obtain and display curves as envelopes of straight lines in a dynamical and interactive way. In a sketch, take two segments AB and CD of equal length, with A and C closer together than B and D. A point F will move on AB and a point E will move on CD. Point F is animated to move with uniform speed starting at A. At the same time E starts at D and moves with the same speed. Using the trace feature of Sketchpad, we will see that the segment FE changes as the points move (see Figure 4). The reader can also see an interactive figure on the web (Figure 2 in Flores [3]). There are other interactive examples of curves as envelopes of families of straight lines or families of circles in this web site.

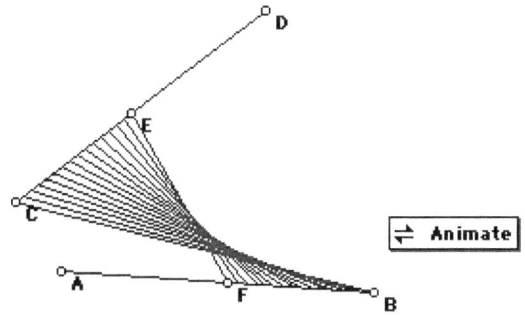

Figure 4. Parabolas with a dynamic geometry program.

Students can also gather some empirical evidence that the envelope does indeed have some of the same properties as the parabola. Figure 5 shows that rays of light reflected at the appropriate point on each line would pass through a single point. The parabola has the property of reflecting parallel rays through the focus of the parabola. This reflective property of the parabola is used for mirrors of telescopes, parabolic antennas, and searchlights.

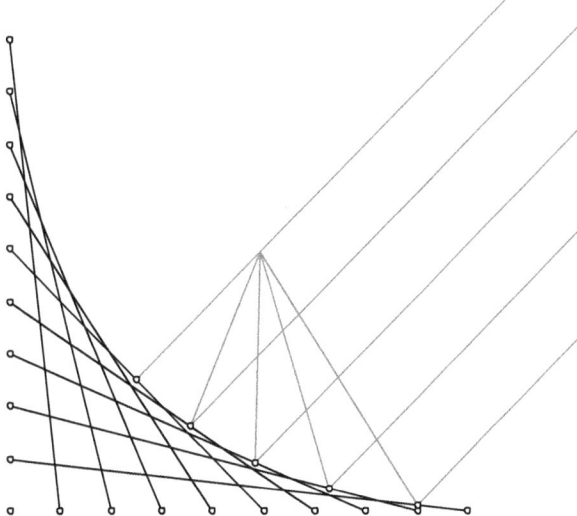

Figure 5. Reflected rays converge at a point.

Empirical evidence is important to help students understand a result and convince themselves that it is true. However, to help our students develop their abilities to prove results in a mathematical way, we need to help them go beyond the empirical schemes of justification and use analytical schemes, that is, the kind of arguments that are of general nature and are considered by mathematicians to be the distinctive way of proving theorems in mathematics (Harel and Sowder [4]).

To facilitate the proof that the generated curve is in fact a parabola we will use analytic geometry to represent the lines. An activity that can help in the transition, and that students like, is to graph the following family of straight lines on a graphing calculator. The calculator draws only straight lines, but students are pleasantly surprised that a curve appears at the end (see Figure 6).

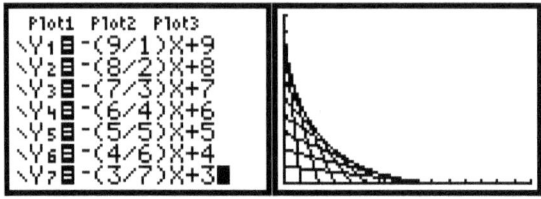

Figure 6. A family of lines on a graphing calculator.

A PROPERTY OF TANGENTS TO THE PARABOLA

Tangent lines to a parabola have the following property. Given three tangent lines to the parabola at A, B, and O, and their points of intersection P, S, and Q (see Figure 7), then the indicated segments satisfy the relation $\frac{SP}{PA} = \frac{QO}{OP} = \frac{BQ}{QS}$ (Wells [10]). In the above example, the axes play a role of SA and SB, and P and Q are chosen so that the relation is satisfied.

Let us see why the relation $\frac{SP}{PA} = \frac{QO}{OP} = \frac{BQ}{QS}$ holds. We prove it for the case $y = x2$; the general case follows using dilations and rigid motions, which preserve ratios (all parabolas are similar in the sense that corresponding parts are proportional). Let the equation that describes the parabola be $y = x2$, and let $A = (x_1, y_1)$, $O = (x_2, y_2)$, $B = (x_3, y_3)$. The derivative of y at the point of tangency A is $2x_1$; the equation for the tangent line is given by $y - x_1{}^2 = 2x_1(x - x_1)$, and similarly for B and O. The equations of the lines through the three points can be written as $y = 2x_1 x - x_1^2$, $y = 2x_2 x - x_2^2$, and $y = 2x_3 x - x_3^2$. Solving pairs of simultaneous equations to find the abscissas for P, S, and Q, we obtain $x_P = \frac{x_1 + x_2}{2}$, $x_S = \frac{x_1 + x_3}{2}$, $x_Q = \frac{x_2 + x_3}{2}$. From here it is easy to see that $\frac{x_S - x_P}{x_P - x_1} = \frac{x_Q - x_2}{x_2 - x_P} = \frac{x_3 - x_Q}{x_Q - x_S}$. Using similar triangles, we can see that this is equivalent to $\frac{SP}{PA} = \frac{QO}{OP} = \frac{BQ}{QS}$.

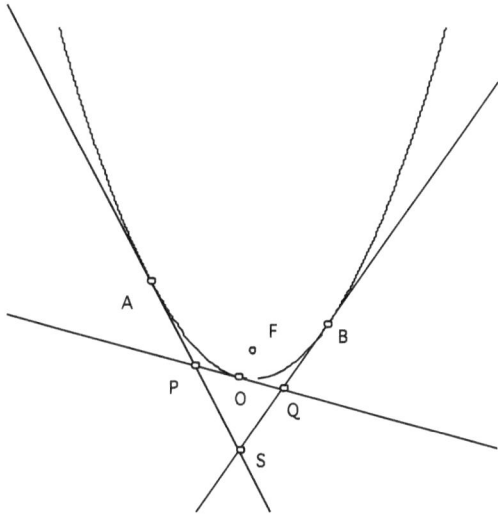

Figure 7. Relationships among three tangents.

THE EQUATION OF THE CURVE TANGENT TO THE STRAIGHT LINES.

To understand better the nature of the curve obtained, let's consider the equations of the lines given in Figure 8. Notice that for each line the sum of the value of the y intercept plus the value of the x intercept is equal to 10. These lines seem to form a curve. As a first step towards proving that the curve that is tangent to all lines is a parabola we will see how we can describe this family of straight lines in terms of a parameter. The equations for these lines are given by $y = -\frac{9}{1}x + 9$, $y = -\frac{8}{2}x + 8$, ..., $y = -\frac{2}{8}x + 2$, $y = -\frac{1}{9}x + 1$. We can describe the family of curves with a parameter p, $y = -\frac{p}{10-p}x + p$, $0 < p < 10$. For each value of p we will get a different line.

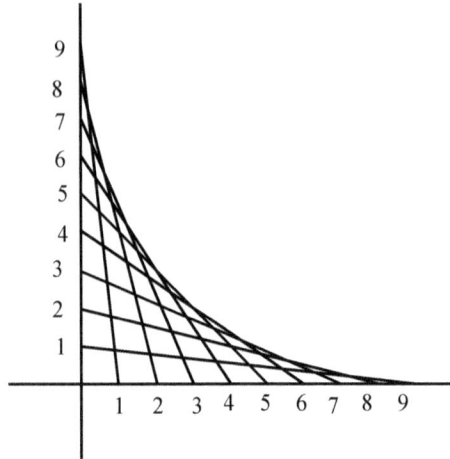

Figure 8. The sum of the x intercept and the y intercept is constant

To find the equation of the curve that is tangent to all the lines in the family, we will apply to the parabola the method described by Boltianskii [1]. Each line can alternatively be described by the equation

$$f(x, y, p) = (10 - p)y + px - p(10 - p) = 0$$

The point where the parabola is tangent to a line with parameter value p is close to the point of intersection of the line with another line of the family that is "close to" the first line, say a line with the parameter value $(p + \varepsilon)$ where ε is a small number.

$$f(x, y, p + \varepsilon) = (10 - (p + \varepsilon))y + (p + \varepsilon)x - (p + \varepsilon)(10 - (p + \varepsilon)) = 0$$

To find the point of intersection, we need to solve the system of equations

1) $f(x, y, p) = (10 - p)y + px - p(10 - p) = 0$

2) $f(x, y, p + \varepsilon) = (10 - (p + \varepsilon))y + (p + \varepsilon)x - (p + \varepsilon)(10 - (p + \varepsilon)) = 0$

However, in order to simplify the computations, rather than dealing with the second equation we will subtract the first from it, and divide by ε. That is, our system of equations will be

1) $f(x, y, p) = 0$

2) $\dfrac{f(x, y, p + \varepsilon) - f(x, y, p)}{\varepsilon} = 0$

We are interested in the limit of the intersection points of the two lines as ε tends to zero.

Notice that the limit of Equation (2) as ε tends to zero is the derivative of f with respect to p.

So our system becomes

1) $f(x, y, p) = 0$

2) $f_p'(x, y, p) = 0$

In our case, $f_p'(x, y, p) = -y + x - 10 + 2p = 0$.

From here, $p = \frac{1}{2}(y - x - 10)$

Substituting this value of p into Equation (1) and simplifying, we obtain

$x^2 - 2xy + y^2 - 20x - 20y + 100 = 0$

We can see that this is the equation of a parabola in different ways. (Some seasoned mathematician could say, just by looking at this equation that it is clear that it is a parabola; an equation of the form $Ax^2 + Bxy + Cy^2 + Dx + Ey + F = 0$ is a parabola if $4AC - B2 = 0$, see [2]. However, this is not part of the trade and parcel of calculus students, nor is such a result commonly found in today's textbooks). One way is using a graphing program such as the graphing calculator you can find on a Macintosh computer. If you use this, try $x^2 - 2xy + y^2 - 20x - 20y + 100 > 0$ and look at the boundary.

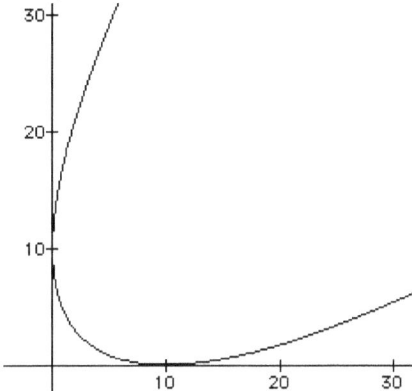

Figure 9. A tilted parabola.

Another way is to change axes. By rotating them 45°, the equation in the new coordinates will be the standard equation of a parabola. For a coordinate rotation of the axes of α degrees, the new coordinates of the point (x, y) are given by

$$\begin{pmatrix} x' \\ y' \end{pmatrix} = \begin{pmatrix} \cos\alpha & -\sin\alpha \\ \sin\alpha & \cos\alpha \end{pmatrix} \begin{pmatrix} x \\ y \end{pmatrix}.$$

Therefore

$$\begin{pmatrix} x \\ y \end{pmatrix} = \begin{pmatrix} \cos\alpha & \sin\alpha \\ -\sin\alpha & \cos\alpha \end{pmatrix} \begin{pmatrix} x' \\ y' \end{pmatrix}$$

For the case $\alpha = 45°$, $\cos 45° = \sin 45° = \dfrac{\sqrt{2}}{2}$, so we can use the substitutions

$$x = \frac{\sqrt{2}}{2}x' + \frac{\sqrt{2}}{2}y'$$

$$y = -\frac{\sqrt{2}}{2}x' + \frac{\sqrt{2}}{2}y'$$

Simplifying the expressions obtained, the new equation (deleting the ') is $10\sqrt{2}y = x^2 + 50$, which clearly is the equation of a parabola.

CONCLUSION

To understand in mathematics is to make connections between ideas, facts, and procedures. According to Hiebert and Carpenter [5], a mathematical idea, procedure, or fact is understood "if its mental representation is part of a network of representations. The degree of understanding is determined by the number and the strength of the connections" (p. 67). All too often students do not have the opportunity to make those connections. We hope that with the kind of activities described here, students will develop a network of related ideas, as well as stronger and more numerous connections. They will have a better understanding of the parabola and some of its properties and applications, as well as of some of the powerful methods and tools that mathematicians have developed over the centuries.

REFERENCES

1. Boltianskii, Vladimir G. 1964. *Envelopes*. New York: Macmillan.
2. Brannan, David A., Espleen, Matthew F. and Gray, Jeremy J. 1999. *Geometry*. Cambridge, England: Cambridge University Press.
3. Flores, Alfinio. 1999. Interactive figures for *A rhythmic approach to geometry*. Available at http://www.public.asu.edu/~aaafp/rhythm.html
4. Harel, Guershon and Sowder, Larry. 1998. Students proof schemes: Results from exploratory studies. *Conference Board of the Mathematical Sciences Issues in Mathematics Education*. 7: 234-283.
5. Hiebert, James and Carpenter, Thomas P. 1992. Learning and teaching with understanding. In Douglas A. Grouws (Ed.), *Handbook of Research on Mathematics Teaching and Learning* (p. 65-97). New York: Macmillan.
6. Jackiw, Nicholas. 1995. *The Geometer's Sketchpad* 3.0 [Computer software]. Berkeley, CA: Key Curriculum Press.
7. Jackiw, Nicholas. 1998. *JavaSketchpad* [on line]. Available http://www.keypress.com/sketchpad/java_gsp/index.html
8. Millington, Jon. 1996. *Curve Stitching*. Norfolk, England: Tarquin Publications.
9. Somervell, Edith L. 1975. *A Rhythmic Approach to Mathematics*. Reston, VA: National Council of Teachers of Mathematics.
10. Wells, David G. 1991. *The Penguin Dictionary of Curious and Interesting Geometry*. London: Penguin Books.

BIOGRAPHICAL SKETCH

Alfinio Flores is Professor of Mathematics Education at Arizona State University. He received his undergraduate and master degrees in mathematics from UNAM in Mexico City. He received his Ph. D. in mathematics education at The Ohio State University. He uses hands-on materials, visual representations, graphing calculators, and computers, to help students understand mathematics better. He enjoys riding his bike home from work.

26 MATHEMATICS OF CHILDREN, MATHEMATICS FOR CHILDREN[26]

One of the most important changes in the new *Principles and Standards* document (National Council of Teachers of Mathematics 2000) is the inclusion of recommendations for pre-kindergarten children. By including younger children into their first age band, the new document makes explicit that students enter school with a wealth of concrete and informal mathematical experiences and knowledge. Furthermore, by including these recommendations with recommendations for school children in the early grades (kindergarten to second grade) as a separate age band. the document also emphasizes that the early grades in school are crucial for students. The idea children will form about school mathematics and how it relates to their own approaches to solve problems will heavily depend to what extent teachers use mathematics of children as their basis of teaching, and to what extent they use mathematics methods and strategies that are appropriate for children at that age.

In the years that range from pre-kindergarten through their first years of formal school, teachers, parents, and caregivers can help develop children's budding mathematical understanding by using what children already know as a basis, rather than conveying the idea that school mathematics does not have a direct relation to children's own methods. Teachers can foster children's informal mathematical abilities and concrete problem solving skills, and at the same time use those approaches to gradually build more powerful and

---

[26] Flores, A. (2000) Mathematics of children, mathematics for children. *New England Mathematics Journal, 32*(2), 18-26. Reprinted by permission of the Association of Teachers of Mathematics in New England.

effective ways to compute and record their thinking. This link to previous knowledge is important. Children understand mathematics if they can see how it is connected or related to other things they know (Hiebert et al. 1997).

Young children solve problems that lend themselves to a mathematical treatment using multiple strategies and approaches. They may act out the problem, use concrete objects such as chips or tokens, use counting strategies, pictorial representations, or mental computation. Most children enter school confident in their abilities to solve problems, and eager to learn mathematics. When students are encouraged to use their methods in school, teachers can build on that positive attitude towards mathematics. Also, recent findings in the science of learning emphasize the importance of active learning and helping students take control over their own learning (Bransford et al. 1999)

Although the numbers that children in the early grades deal with are generally small, children often reveal a deep understanding of fundamental number properties in the strategies they use. For example, by breaking the number 7 into $2 + 5$, to solve a problem such as $8 + 7$, saying 8 and 2 is ten, and 5 is 15, therefore $8 + 7 = 15$, students reveal understanding of the associative property for addition. By being excited about the fact that $2 + 3$ is the same as $3 + 2$, a child reveals his understanding of the commutative property and that it applies when adding any two numbers.

Of course, at this level, there is no need to use the technical terms like "associative" or "commutative", but by using and emphasizing these properties of numbers and operations, children start developing also thinking of algebraic nature.

Children's ways of thinking often also reveal their understanding of the value of the numbers involved. They break numbers into smaller numbers that are easier to deal with; often the smaller numbers correspond to groupings of ten, which can be used to develop their understanding of our number system, in particular place value. For example, when adding numbers like $24 + 18$, children frequently add the tens first, "$20 + 10$ is 30, $4 + 8$ is 12, $30 + 12$ is 42." When working with concrete objects that represent the numbers, such as rods of cubes that represent tens and individual cubes that represent units, children frequently deal with the bigger objects first. The teacher can help students develop a way to record their thinking and their actions that reflects what students actually did.

In the problem above, the teacher may introduce a notation that reflects the method of the student and emphasizes place value, for example (figure

$$
\begin{array}{r}
24 \\
+\ 18 \\
\hline
30 \\
+\ 12 \\
\hline
42
\end{array}
$$

Figure 1. Recording student's process

Writing not only the answer, but also the intermediate steps in the procedure has two advantages. On one hand, it is a record that allows students to reflect on their own thinking. On the other hand, the thinking of the student can be brought into the forum of discussion and analysis. In addition, by using the written record students will develop ways that will make their thinking more efficient when they work with bigger numbers or in problems that require multiple steps.

Young children also use their own ways of representing quantities using drawings or other representations that may not be clear to adults. Their use of symbols may not be consistent with the conventional ways of using the symbols. Nevertheless, those representations constitute children's efforts to record or communicate their thinking. Rather than just dismiss children's ways, because the answer is incorrect, the teacher can talk to the student to understand the thinking of the child, and use the child's representation as a way to introduce or explain the notation commonly used. For example, Gordon, a first grader, who already understood what one half meant, was recently introduced to the notation $\frac{1}{2}$. For the problem $\frac{1}{2} + \frac{1}{2} + \frac{1}{2}$ he obtained the correct answer, three halves. To record the answer he wrote $3\frac{1}{2}$, which was consistent with ways he used to write answers like 3 apples, 3 cookies, but that is not consistent with the way we conventionally identify $3\frac{1}{2}$ with the meaning $3+\frac{1}{2}$ rather than with the meaning $3\times\frac{1}{2}$. By talking to the child the teacher can find out what he means, and share with him accepted conventional notations to represent the correct answer.

Sometimes children's methods can be quite elaborate and involve keeping track in their heads of several partial answers. Consider for example the case of Luis, a second grader who had not yet learned multiplication in school. To solve the problem 7 x 5, Luis added silently in his head and obtained the answer 35. When asked how he had done it, he explained: seven and three is ten, and four is fourteen; and six is twenty and one is twenty one; and seven is twenty eight; and two is thirty and five is thirty five. Luis' method to add 7 + 7 + 7 + 7 + 7 revealed that he was fluent in breaking numbers to form tens, and that he was able to keep track at the same time how many sevens he had added. However, such a method becomes too complex and prone to errors when the numbers become bigger. The teacher can help students build on their methods and understandings, and use properties of numbers and operations, and the facts they already know to become more efficient.

When dealing with bigger numbers, students can use related facts with numbers close to the one they are interested. For example, another child used the following strategy to multiply 12×3. The child knew how to multiply by ten and by eleven. He said "$11 \times 3 = 33$, and 3 more is 36." The child revealed not only an understanding of multiplication as repeated addition, but also used implicitly the distributive property of multiplication over addition. $12 \times 3 = (11 + 1) \times 3 = 33 + 3 = 36$.

The teacher can encourage children to use groups of numbers so that they do not have to perform so many individual computations, and so that at the same time, keeping track of the intermediate steps becomes easier. When they want to solve a problem that involves adding 9, rather than counting on nine units, students frequently use adding ten as an intermediate step. To solve $17 + 9$ they may say, 17 plus ten is 27, one less is 26.

Many students in the early grades use the strategy of counting on to solve addition or subtraction problems. This method can be a very effective strategy if the number added is not too big, or in the case of subtraction, if the difference is not too big. However, it can become cumbersome for bigger numbers. The teacher can help students develop more efficient ways to count on, for example by counting on by tens or even hundreds. Thus a student who wants to add $24 + 32$ could say, 24 (pause) 34, 44, 54, 55, 56.

Contexts are important to develop meaning for different operations and their relations to other operations. Children's strategies often vary according to the context. For example, children can compare a collection of 3 objects with another with 5 objects, to find the difference. Or they can find out how many objects are left if they start with five and give away 3. Or they may want to find out how many objects they need if they want to have a collection of five but so far they only have three. Students can use different strategies such as adding on objects, counting on numbers, counting backwards, or modeling a collection with concrete objects and taking away three. Story problems often provide a context and help students understand the meaning of operations.

It represents a huge step for many children to realize that all those problems can be represented by 5 - 3. It is therefore important that the symbolic notation is not introduced prematurely. Also, it is important that students do not develop a too restricted interpretation of 5 - 3. At the same time that they solve the original problems, students can make the connection to the related addition problem 3 + 2 = 5.

Together with the number standard, geometry is the other standard that receives a major emphasis in the early grades. At an early age, children's ability to recognize and differentiate shapes is often more developed than their number skills. Young children may be able to see that a square and an equilateral triangle are different by considering them as a whole, even when they cannot count the numbers of sides in a square or a triangle.

The teacher can help the student learn concepts such as triangles and quadrilaterals by providing a rich sample of examples and also non examples. It is very common that young children are only shown one kind of triangle (equilateral), in only on position. By having triangles that have different shapes (see figure 2) and that can be manipulated or are drawn in different positions, children will be able to form a concept image of triangle that is not too restricted.

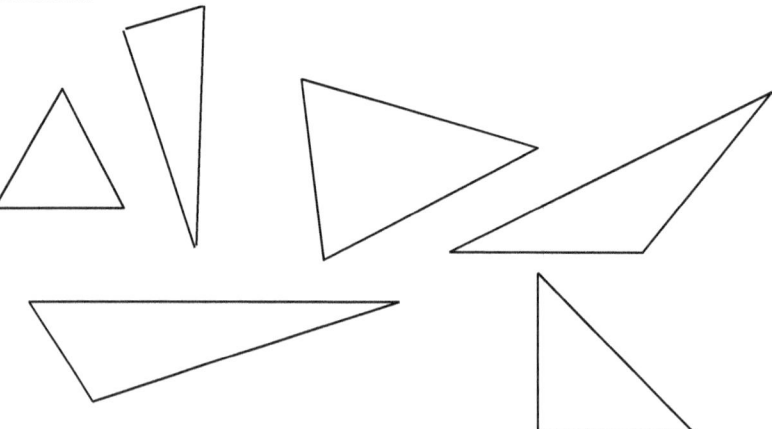

Figure 2. Examples of triangles

The role of non examples is also important to clarify and make their concept more precise. When non examples are not discussed, some children may identify as triangles shapes that have a base and a point pointing up (see figure 3).

Figure 3. Non examples of triangles

An important step in the development of geometric thinking is the transition from focusing on the shapes as a whole to considering also their parts. Teachers can provide students in the early grades with opportunities to focus on parts of the shapes, such as the sides of a square and its angles. Often, children's language does not clearly reveal their thinking, and the teacher may have to observe what the child does, or may have to ask additional questions to understand what the child means, in terms of the child's thinking and not just from the adult perspective.

Children's vocabulary at this age band is still growing at a rapid pace. Teachers can use the informal language of the child as a base to introduce mathematical terms that will help students express their thinking more precisely. Consider the following discovery done by a kindergarten student working with wooden blocks of different shapes. The child stated that a square was equal to a triangle, and used a block with a square face, another with a triangular face and a block with a rectangular face that was equal to two of the same squares and two of the same triangles to demonstrate his statement. He showed that both the block with the square face and the block with the triangular face where half of the larger block (figure 4), and thus had to be equal.

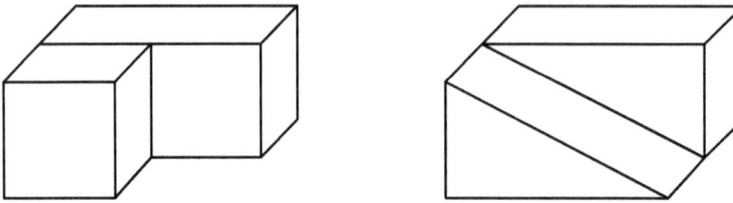

Figure 4. Each smaller block is one half of the bigger block.

By dialoguing with the student the teacher can find out what the student meant by "equal" and understand whether the child was referring to the surface of the face of the blocks or to the blocks themselves, and thinking in terms of volume. The teacher can help the student make his statement more precise.

Teachers play a crucial role in the early grades to use children's informal and concrete knowledge of mathematics as they enter school as the basis for learning school mathematics, and provide gradual guidance to more powerful and general methods that are appropriate for children.

References

Bransford, J. D., Brown, A. L., and Cocking, R. R. (Eds.). *How people learn: Brain, mind, experience, and school.* Washington, DC: National Academy Press, 1999.

Hiebert, J. et al. *Making sense: Teaching and learning mathematics with understanding.* Portsmouth, NH: Heinemann, 1997.

National Council of Teachers of Mathematics. *Principles and standards for school mathematics.* Reston, VA: National Council of Teachers of Mathematics, 2000.

27 MECHANICAL ARGUMENTS IN GEOMETRY[27]

ABSTRACT: Concepts of mechanics such as velocity; composition, resultants, and equilibrium of forces; potential energy; center of gravity; and moment of force are used to give alternative proofs of geometrical results.

KEYWORDS: Mechanics, geometry, integration of mathematics and physics, interactive geometry, dynamical geometry.

INTRODUCTION

In this paper we present mechanical arguments to prove results in geometry. We use concepts of physics such as velocity; composition, resultants, and equilibrium of forces; potential energy; center of gravity; and moment of force. The examples given here can be used in physics courses to give students the opportunity to apply mechanical concepts to a different field. On the other hand, the students in mathematics will have an opportunity to see alternative and highly intuitive proofs for results in geometry. In addition, a dynamical geometry program [8] and its world wide web component [7] were used to develop interactive figures to accompany this paper. As the reader goes through the text, he or she can also interact with some of the figures in a dynamical environment in the corresponding web site [3].

AN ARGUMENT OF VELOCITY

---

[27] Flores, A. (1999). Mechanical arguments in geometry. *PRIMUS, 9*(3), 241 - 250. Reprinted by permission of Taylor & Francis (http://www.tandfonline.com).

Let ABC be any triangle, and choose any point D. Points D_1, D_2, and D_3 are obtained the following way: D_1 is the reflection of D around B, D_2 is the reflection of D_1 around A, and D_3 is the reflection of D_2 around C. What can you say about the position of the midpoint M between D and D_3? [1].

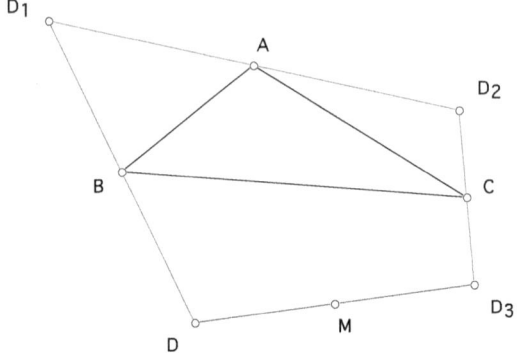

Figure 1. Midpoint M

The reader may draw a triangle, then choose D in different positions, follow the instructions to obtain M (see also interactive Figure 1), and observe that the position of M does not depend on the initial position of D. A form to convince ourselves that this is the case is the following. If point D is moving with velocity **v**, point D_1 will move with a velocity of the same magnitude but opposite direction **-v**, because the segment BD is always congruent to segment BD_1, but the two segments change in opposite directions. At the same time, point D_2 moves in opposite way with respect to D_1, so that the velocity of D_2 is also **v**. At the same tine D_3 moves with a velocity of the same magnitude but opposite direction to the velocity of D_2, that is, the velocity of D_3 is **-v**. Thus, if D moves in any way, D3 moves exactly in opposite direction (see fig. 2, or interactive figure 2). Therefore, the velocity of the midpoint M, which is the average of the velocities of D and D_3, $(-\mathbf{v} + \mathbf{v})/2$, is always equal to zero. Therefore, point M does not move, thus being independent of the initial position of D. For more examples of how geometric results about a geometric figure can be obtained from considering the velocity of endpoints of segments in changing figures using a dynamical geometry program, see [4].

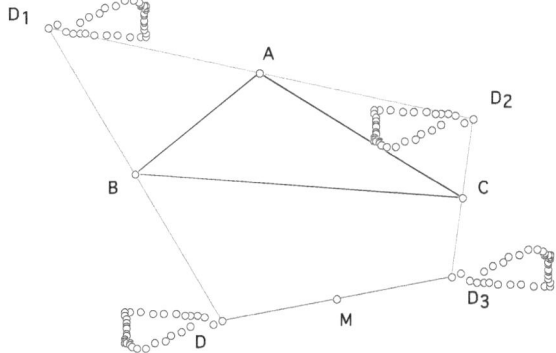

Figure 2. Opposite velocities

COMPOSITION OF FORCES

Kogan [9] uses arguments based on the composition of forces acting along the sides of triangles to prove that certain segments intersect in a single point. The method is based on the basic fact that if three forces are in equilibrium, then the line of action of these forces intersect in a single point. Let's see how the method is used to show that the angle bisectors of a triangle all meet in one point. Consider six forces of the same magnitude, F_1, F_2, ..., F_6 acting along the sides of a triangle, as shown in figure 3. The resulting forces R_{16}, R_{23} and R_{45} are directed along the angle bisectors of the interior angles of the triangle, because each parallelogram is a rhombus, and the diagonal is the angle bisector. Because the forces F_i cancel each other in pairs, they are in equilibrium. Therefore, the resultants R_{16}, R_{23} and R_{45} are also in equilibrium, so that the angle bisectors intersect in a single point.

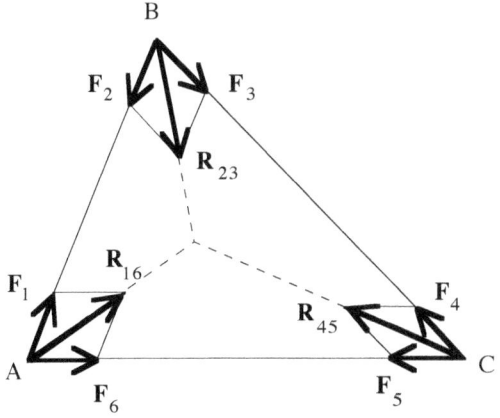

Figure 3. Forces in equilibrium.

MINIMUM DISTANCE TO THE THREE VERTICES OF A TRIANGLE

Join point P inside the triangle with the three vertices (fig. 4, or interactive figure 4). We want to find the location of P so that the sum of the distances PA + PB + PC be minimum.

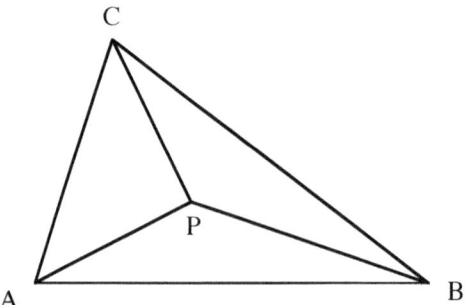

Figure 4. Distances to three vertices

One way to solve this problem is with the following mechanical argument suggested by De Finetti [2]. Consider the triangle rigid but weightless in horizontal position. Through holes on the vertices of the triangle we pass a string with equal masses attached to each one (the strings move freely through the holes). Bind the three strings together at the movable point P (fig. 5).

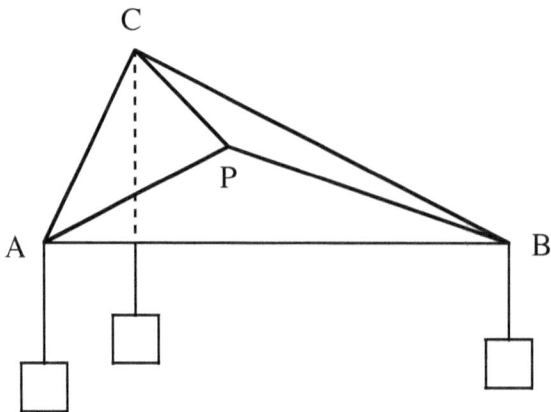

Figure 5. Equal masses hanging.

The system of masses will be in equilibrium when the potential energy is minimal. This means that the sum of the distances from the vertices to the masses is maximal. Because the total length of the strings is constant, this in turn means that AP + BP + CP is a minimum. Because the three forces acting along AP, BP, and CP are equal, the resultant of two of them is on the angle bisector between them, and because the resultant is in equilibrium with the third force, the line of action of this one is also the angle bisector. Therefore, each of the segments PA, PB, y PC is on the angle bisector of the other two segments. From here it follows that each pair of segments form an angle of 120° at P (fig. 6). Polya [10] uses the same kind of argument, but with the triangle in vertical position. For other geometrical results obtained from considerations made from attaching masses to different elements of triangles, see [5].

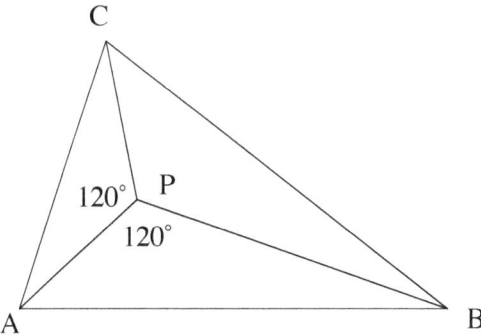

Figure 6. Minimum distance to three vertices.

SUM OF TRIANGULAR AREAS IN A RECTANGLE

A rectangle is escribed around an equilateral triangle ABC in any direction as shown in fig. 7. In general, each side of ABC cuts the a right triangle from the rectangle. Prove that the sum of the two areas of the smaller right triangles is equal to the area of the bigger right triangle (see interactive figure 7).

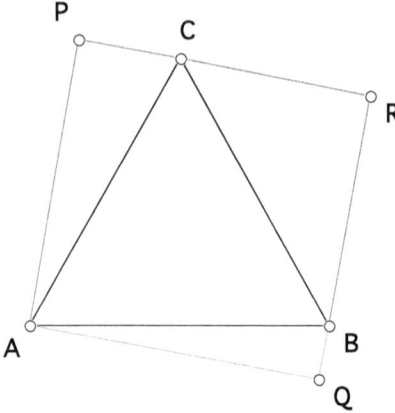

Figure 7. Sum of areas.

The following solution was found by Mrs. Dijkstra when she was octogenarian [6]. Rotate triangle CPA 60° (counter clock wise) around C so that CA coincides with CB. Rotate triangle BQA 60° (clock wise) around B so that AB coincides with CB (see fig. 8).

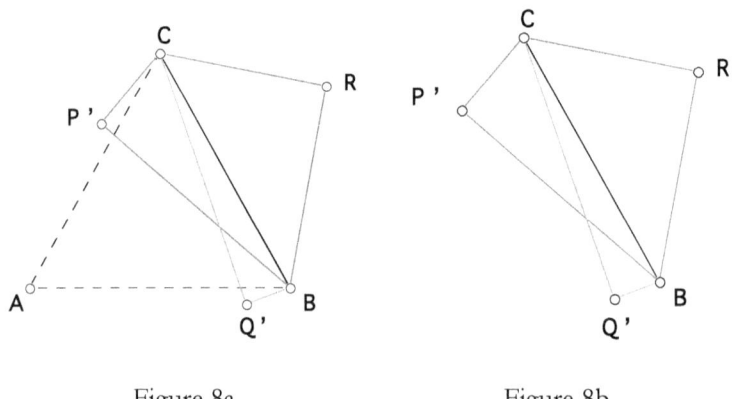

Figure 8a   Figure 8b

Because the angles at P', Q' and R are right angles, the circle with diameter CB will pass through these three points (fig. 9a). Angle CRP' is 120° (why?), and also angle RBQ' is 120°. Therefore points R, P', and Q' are the vertices of an equilateral triangle (fig. 9b).

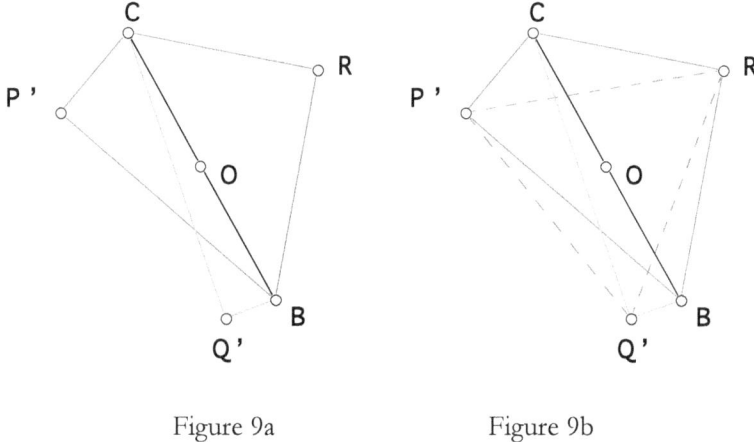

Figure 9a   Figure 9b

If we hang equal masses from the vertices R, P', y Q', the system will have its center of gravity at the center O of the circle. Because O is on the diameter CB, the system will be balanced with respect to the diameter CB. Therefore, the moments of force around CB generated by the masses on P' and Q' are equilibrated by the moment of force of the mass at R (fig 10a).

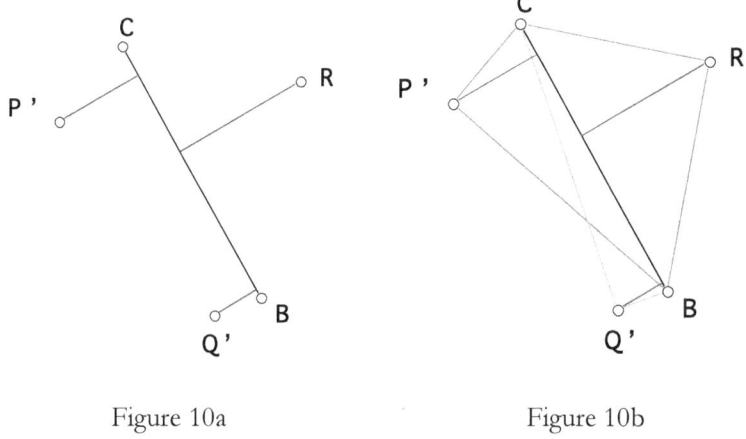

Figure 10a   Figure 10b

Therefore, the sum of the heights of the triangles CP'B and CQ'B is equal to the height of triangle CRB (fig. 10b). Thus the sum of the areas of the two smaller triangles is equal to the area of the bigger triangle.

Dijkstra uses for the particular case of the equilateral triangle a general principle that can be stated in the following way. Let ABC be any triangle. Let G be the point of intersection of the medians, Let *l* be an arbitrary line through G (fig. 11, also interactive fig. 11). Then the sum of the two shorter distances from two vertices to the line is equal to the longer distance from the other vertex. To convince ourselves that this is the case, imagine the triangle being rigid but without mass. Put equal masses on each vertex. The center of gravity of the system is G, therefore the system will be balanced on G, or on any line through G. Therefore the sum of the moments of force on one side of the line is equal to the sum of the moments on the other side $md_1 + md_2 = md_3$, therefore $d_1 + d_2 = d_3$.

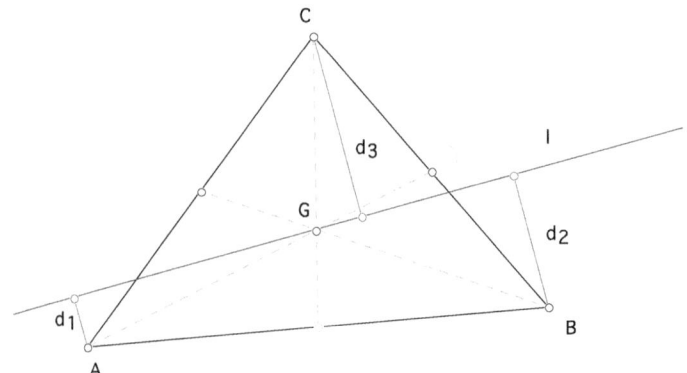

Figure 11. Distances to line.

CONCLUSION

Students frequently use mathematics, and specifically geometry to solve and prove problems in physics. However, they are seldom aware that they can use concepts in physics, in particular in mechanics to prove theorems in mathematics. As was shown in this paper, physics and mathematics can be of mutual service. The two-way interaction between mechanics and geometry will without doubt enrich the learning of the students in both fields.

REFERENCES

1. Balacheff, N. (1998). Learning mathematics as modelling. Paper presented at the Technology and Standards 2000 Conference. Washington, DC (June).
2. De Finetti, B. (1974). <u>Die Kunst des Sehens in der Mathematik</u>. Basel, Birkhäuser.
3. Flores, A. (1998) Mechanical arguments in geometry [On line]. Available http://www.public.asu.edu/~aaafp/mechanicalarguments.html
4. Flores, A. (1998). The kinematic method and the Geometer's Sketchpad

in geometrical problems. <u>International Journal of Computers for Mathematical Learning</u>, <u>3</u>, 1-12.

5. Flores Peñafiel, A. (Submitted). Triángulos pesados. <u>Miscelánea Matemática</u>.

6. Honsberger, Ross. (1985). <u>Mathematical gems 3</u>. Mathematical Association of America.

7. Jackiw, N. (1998). JavaSketchpad [On line]. Available http://www.keypress.com/sketchpad/java_gsp/index.html

8. Jackiw, N. (1995). The Geometer's Sketchpad 3.0 [Computer software]. Berkeley, CA: Key Curriculum Press.

9. Kogan, B. Yu. <u>The applications of mechanics to geometry</u>. University of Chicago Press, 1974.

10. Pólya, G.. (1954). <u>Mathematics and plausible reasoning. Vol. 1 Induction and analogy in mathematics</u>. Princeton, NJ: Princeton University Press.

BIOGRAPHICAL SKETCH

Alfinio Flores is now a seasoned (mainly due to the hot weather) professor of mathematics education at Arizona State University. One of his interests is to look for alternative ways to make geometry more accessible for students and teachers. His approaches include the use of concrete objects, stories and fairy tales, dynamical geometry software, other computer programs, graphing calculators, and ideas from physics and nature. He is also fond of finding geometrical representations of numerical and algebraic identities.

28 THE LAW OF COSINES: CONNECTIONS FOR FUTURE TEACHERS[28]

ABSTRACT: Prospective secondary teachers can explore mathematics at their own level by using the law of cosines to establish connections between topics that are usually taught separately, such as cosine of the difference of two angles, Cauchy's inequality, determinants, sine of the difference of two angles, triangle inequality, and inner product of two vectors.

KEYWORDS: Law of cosines, trigonometric identities, vectors, determinants, inequalities, mathematical connections.

INTRODUCTION

A recommendation for the mathematical preparation of teachers is that they should have an understanding of mathematics beyond of what they are going to teach. A common practice is to require prospective secondary teachers to take more advanced and abstract mathematics courses, so that they can have a view of elementary mathematics from an advanced standpoint. However, frequently missing from their preparation is an understanding of connections between topics at the same level of abstraction as the material that they are going to teach. Developing a network of concepts is important for two reasons. First, prospective teachers can explore and investigate on their own by using the mathematical tools and concepts that are within their reach. Second, by knowing how different concepts are connected, when they become teachers, they will be able to guide their students to explore

---

[28] Flores, A. (1999). The law of cosines: Connections for future teachers. *PRIMUS*, *9*(2), 123-132. Reprinted by permission of Taylor & Francis (http://www.tandfonline.com).

mathematics at the students' own level. This will also help improve students' understanding of mathematical concepts as they use and apply them in different contexts.

Steffe [7] suggests that the law of cosines can be used to give unity to several topics in secondary school mathematics. It can also be used in college courses as an opportunity for prospective secondary teachers to engage in open explorations. The law of cosines connects concepts that are usually learned with no relation to each other. College students can explore these connections as part of the courses where the topics such as the cosine of the difference of two angles, Cauchy's inequality, determinants, sine of the difference of two angles, triangle inequality, and inner product of two vectors are taught or used.

Another place where this material can be used is in a course designed to help students in the transition towards courses where the main emphasis is to solve problems and prove results. By solving problems and establishing connections between topics they are familiar with, they will develop strategies that will be helpful to solve problems and establish connections in topics that are new to them. Through these kinds of experiences, future mathematics teachers can "develop a sense of self confidence and a willingness to explore and learn new mathematics on their own" (Leitzel [4], p. 2).

Students can engage in discovery activities with more or less guidance. Rather than presenting the material described in this paper to the student, the instructor may want to use this article as a source of problems (and possible hints) for the students. For example, the instructor may just ask students to prove results or find connections (using the law of cosines or otherwise), or may give them a suggestive figure and additional hints. Students, of course, may derive the connections in ways that are different than those presented here. The instructor may encourage alternative solutions, and students can share different ways of reaching the result.

LAW OF COSINES

The law of cosines $c^2 = a^2 + b^2 - 2ab \cos\gamma$ is a generalization of the Pythagorean theorem. It can be derived from it as shown by Sipka [6] (figure 1).

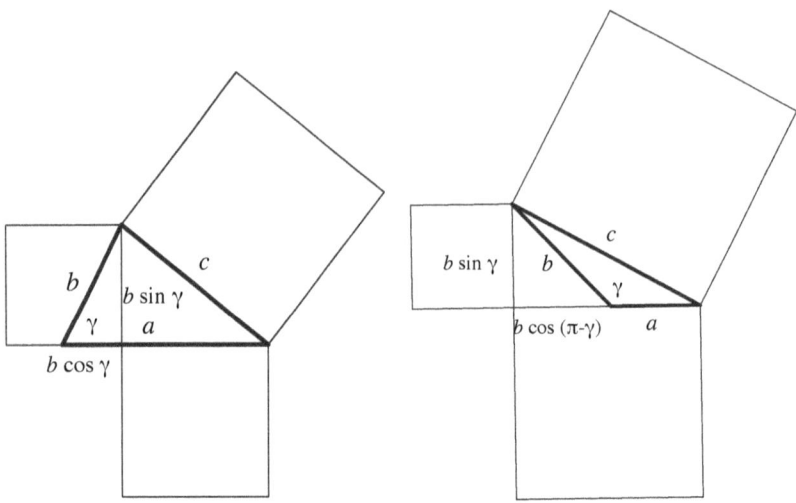

Figure 1. A proof based on the Pythagorean theorem.

$$c^2 = (b\sin\gamma)^2 + (a - b\cos\gamma)^2 = b^2\sin^2\gamma + a^2 - 2ab\cos\gamma + b^2\cos^2\gamma$$
$$= a^2 + b^2 - 2ab\cos\gamma$$

Figure 2 suggests another way to prove the law of cosines.

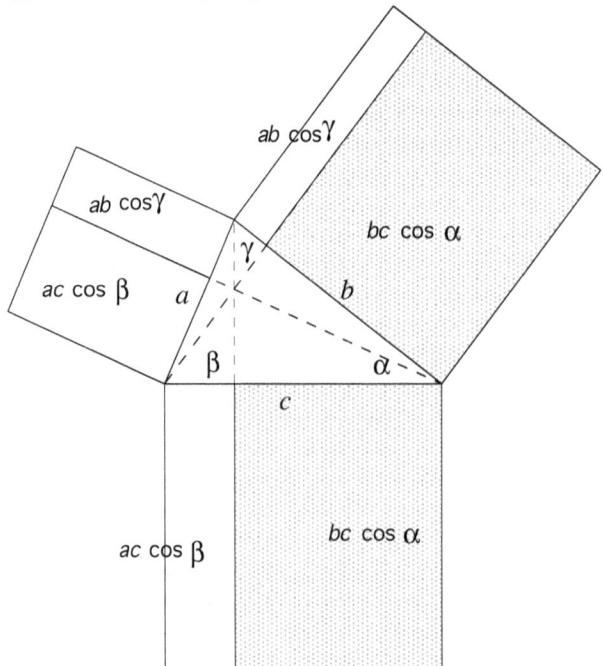

Figure 2. A proof based on equal areas.

LAW OF COSINES AND THE COSINE OF THE DIFFERENCE OF TWO ANGLES.

Draw angles α, β on the unit circle. Let d be the length of the segment that connects the endpoint of α with the endpoint of β (see figure 3).

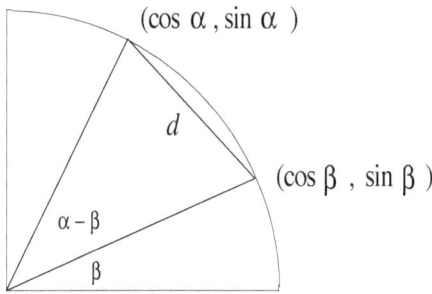

Figure 3. Cosine of the difference of two angles.

On one hand, the distance between points is $d^2 = (\cos\alpha - \cos\beta)^2 + (\sin\alpha - \sin\beta)^2$. On the other, using the law of cosines $d^2 = 1 + 1 - 2\cos(\alpha - \beta)$. From here we have $\cos(\alpha - \beta) = \cos\alpha\cos\beta + \sin\alpha\sin\beta$.

LAW OF COSINES AND CAUCHY'S INEQUALITY

The law of cosines is also related to fundamental inequalities (Beckenbach & Bellman [1]). Consider the triangle OPQ shown in figure 4.

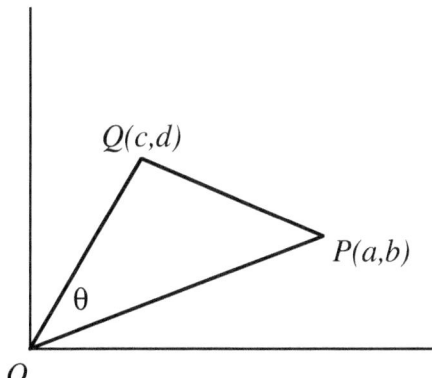

Figure 4. $(a-c)^2 + (b-d)^2 = a^2 + b^2 + c^2 + d^2 - 2\sqrt{a^2+b^2}\sqrt{c^2+d^2}\cos\theta.$

By the law of cosines, $(a - c)^2 + (b - d)^2 = a^2 + c^2 + b^2 + d^2 - 2\sqrt{a^2 + b^2}\sqrt{c^2 + d^2}\cos\theta$. Therefore $\cos\theta = \frac{ac+bd}{\sqrt{a^2+b^2}\sqrt{c^2+d^2}}$. Squaring both sides we obtain $\cos^2\theta = \frac{(ac+bd)^2}{(a^2+b^2)(c^2+d^2)}$. Because $\cos^2\theta \le 1$, (1) $(a^2 + b^2)(c^2 + d^2) \ge (ac + bd)^2$, known as Cauchy's inequality.

Another way to obtain this result is by noticing that

(2) $(a^2 + b^2)(c^2 + d^2) = (ac + bd)^2 + (ad - bc)^2$. Because the square $(ad - bc)^2 \ge 0$, we get (1). Equality in (1) when $ad - bc = 0$. If $b \neq 0$ and $d \neq 0$, this means that $a/b = c/d$. We will give a geometrical interpretation of this in the next section.

CAUCHY'S INEQUALITY AND THE GEOMETRIC INTERPRETATION OF THE DETERMINANT.

Dividing both sides of (2) by $(a^2 + b^2)(c^2 + d^2)$ we obtain

(3) $\quad 1 = \frac{(ac+bd)^2}{(a^2+b^2)(c^2+d^2)} + \frac{(ad-bc)^2}{(a^2+b^2)(c^2+d^2)}$

The first term of the right side of (3) is $\cos^2\theta$, the second term has to be $\sin^2\theta$, because $\cos^2\theta + \sin^2\theta = 1$. Therefore, $\sin^2\theta = \frac{(ad-bc)^2}{(a^2+b^2)(c^2+d^2)}$.

We can rewrite this as $(ad - bc)^2 = (a^2 + b^2)(c^2 + d^2)\sin^2\theta$ or $|ad - bc| = (OP)(OQ)|\sin\theta|$. The right side is the area of the parallelogram $OPRQ$ (see figure 5), and the left side is the absolute value of the determinant of the matrix $\begin{pmatrix} a & c \\ b & d \end{pmatrix}$. The area is zero if O, P, and Q are on the same line, that is, if one column of the matrix is a multiple of the other. We can think the matrix as representing a linear transformation (with the canonical basis of $\mathbf{R}^2$). The images of the basis vectors are the columns of the matrix $\begin{pmatrix} a & c \\ b & d \end{pmatrix}, \begin{pmatrix} a & c \\ b & d \end{pmatrix}\begin{pmatrix} 1 \\ 0 \end{pmatrix} = \begin{pmatrix} a \\ b \end{pmatrix}$, and $\begin{pmatrix} a & c \\ b & d \end{pmatrix}\begin{pmatrix} 0 \\ 1 \end{pmatrix} = \begin{pmatrix} c \\ d \end{pmatrix}$. If $b \neq 0$, and $d \neq 0$, he two vectors will be on the same line if $a/b = c/d$ and the determinant will be zero.

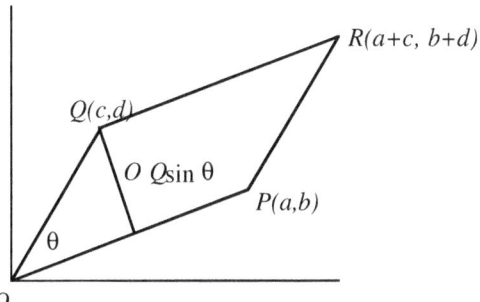

Figure 5. The determinant as an area.

SINE OF THE DIFFERENCE OF TWO ANGLES.

When points P and Q are on the unit circle (figure 6), with coordinates $(\cos \alpha, \sin \alpha)$ and $(\cos \beta, \sin \beta)$, the above result yields $\sin(\alpha - \beta) = \sin \alpha \cos \beta - \cos \alpha \sin \beta$.

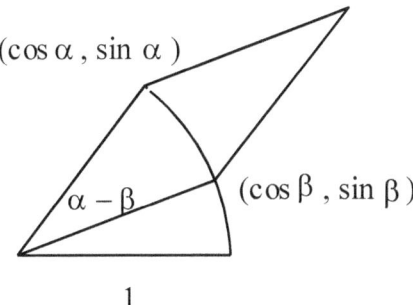

Figure 6. Sine of the difference of two angles.

TRIANGLE INEQUALITY AND CAUCHY'S INEQUALITY

In a triangle the sum of the lengths of two sides is equal to or greater than the length of the third side. Let $R = (a + c, b + d)$. Then $OP + PR \geq OR$ (figure 7) This is equivalent to the inequality $\sqrt{a^2 + b^2}\sqrt{c^2 + d^2} \geq \sqrt{(a + c)^2 + (b + d)^2}$. By squaring both sides and canceling, we see that it is equivalent to $\sqrt{a^2 + b^2}\sqrt{c^2 + d^2} \geq ac + bd$, which is a consequence of the Cauchy inequality.

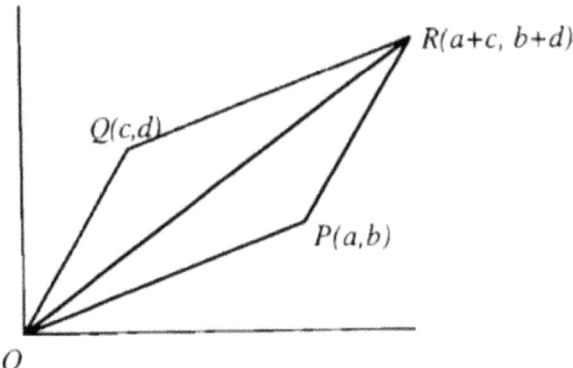

Figure 7. Triangle inequality.

THE LAW OF COSINES AND THE INNER PRODUCT OF TWO VECTORS.

We define the projection a_c of a vector $\boldsymbol{a}$ on a vector $\boldsymbol{c}$ as $a_c = \|\boldsymbol{a}\|u_c \cos\theta$ where θ is the angle between the two vectors, and u_c is a unit vector in the direction of $\boldsymbol{c}$. The projection of $\boldsymbol{c}$ on $\boldsymbol{a}$ is given by $c_a = \|\boldsymbol{c}\|u_a \cos\theta$. The relation between a_c and c_a is given by $\frac{a_c}{\|a\|} = \frac{c_a}{\|c\|}$ (see figure 8). Therefore $a_c\|\boldsymbol{c}\| = c_a\|\boldsymbol{a}\| = \|\boldsymbol{a}\|\|\boldsymbol{c}\| \cos\theta$. The number $\|\boldsymbol{a}\|\|\boldsymbol{c}\| \cos\theta$ is called the inner product of vectors $\boldsymbol{a}$ and $\boldsymbol{c}$ (Friedrichs [3]). It is also called dot product and denoted by $a \cdot c$.

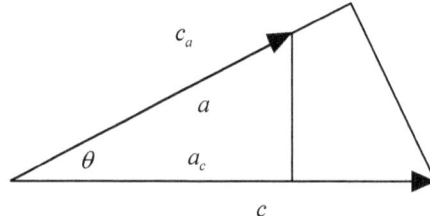

Figure 8. Projections of vectors.

The law of cosines can be used to express the inner product of two vectors in terms of their coordinates. On one hand $\|X - Y\|^2 = \|X\|^2 + \|Y\|^2 - 2\|X\|\|Y\| \cos\alpha$ (figure 9).

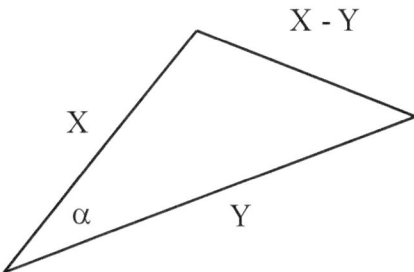

Figure 9. Difference of two vectors.

On the other hand, $\|X - Y\|^2 = (x_1 - y_1)^2 + (x_2 - y_2)^2$. Expanding the right sides and canceling, we obtain $\|X\|\|Y\| \cos \alpha = x_1 y_1 + x_2 y_2$. (In many books, the inner product of two vectors $X = (x_1, x_2)$ and $Y = (y_1, y_2)$ is defined by $(x_1, x_2) \cdot (y_1, y_2) = x_1 y_1 + x_2 y_2$, an approach that makes many students wonder where this formula came from.)

We can also state Cauchy's inequality as $\|X\|\|Y\| \geq |x_1 y_1 + x_2 y_2| = |X \cdot Y|$. Also, the law of cosines can be expressed in terms of the inner product,

$$\|X - Y\|^2 = (X - Y) \cdot (X - Y) = X \cdot X + Y \cdot Y - 2X \cdot Y$$
$$= \|X\|^2 + \|Y\|^2 - 2X \cdot Y$$

In particular, the projection of a unit vector $(\cos \beta, \sin \beta)$ on another unit vector $(\cos \alpha, \sin \alpha)$ is given by $(\cos \alpha, \sin \alpha) \cdot (\cos \beta, \sin \beta)$. The projection is also equal to $\cos(\alpha - \beta)$ (see figure 10). Thus,
$\cos(\alpha - \beta) = \cos \alpha \cos \beta + \sin \alpha \sin \beta$.

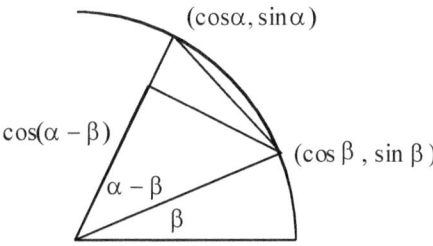

Figure 10. Projection of unit vector.

CONCLUSION

Prospective high school teachers can find connections between the law of cosines and other important concepts in secondary school and freshman college mathematics. They can find alternative proofs of this law. They can find connections to fundamental inequalities such as Cauchy's inequality, and the triangle inequality. An interpretation of a 2 by 2 determinant as the area of a parallelogram is also related to these results, as well as the formulas for the cosine and sine of a difference of angles. The law of cosines is also related to the inner product of two vectors.

Letting prospective teachers find alternative ways to derive the results will also provide opportunities for them to discover and explore. The book by Nelsen [5] can be a source of alternative ideas. Prospective teachers can also explore how other basic results, such as the law of sines can be related to trigonometric formulas such as the sine of the sum of two angles (see for example Coxeter [2]). The possibilities for exploration and discovery for prospective secondary mathematics teachers and their students are unending. Let us provide future teachers with such opportunities, so that in turn they provide their own students with similar opportunities.

REFERENCES

1. Beckenbach, E. & Bellman, R. (1961). *An introduction to inequalities.* New York: Random House.
2. Coxeter, H. S. M. & Greitzer, S. L. (1967) *Geometry revisited.* New York: Random House.
3. Friedrichs, K. O. (1965). *From Pythagoras to Einstein.* Washington, DC: Mathematical Association of America.
4. Leitzel, J. R. C. (Ed.). (1991). *A call for change: Recommendations for the mathematical preparation of teachers of mathematics.* Washington, DC: Mathematical Association of America.
5. Nelsen, R. B. (1993). *Proofs without words.* Washington, DC: Mathematical Association of America.
6. Sipka, T. A. (1988). Law of cosines: Proof without words. *Mathematics Magazine, 61,* 113 (also in Nelsen, 1993, p. 31).
7. Steffe, L. P. (1990). On the knowledge of mathematics teachers. In Davis, R. B., Maher, C. A. & Noddings, N. (Eds.), *Constructivist views on the teaching and learning of mathematics* (p. 167-184). Reston, VA: National Council of Teachers of Mathematics.

BIOGRAPHICAL SKETCH

Alfinio Flores teaches mathematics *con ganas* to prospective teachers at Arizona State University. He uses different approaches to make mathematical abstractions meaningful to his students. They use concrete materials, write and talk about mathematics, use computers and calculators, and make connections with other abstractions. He received his training as a mathematician at UNAM in Mexico, and in mathematics education at The Ohio State University.

ABOUT THE AUTHOR

Alfinio Flores is professor of Mathematics Education at the University of Delaware. He has Mathematics degrees from the National University in Mexico and in Mathematics Education from Ohio State University. He teaches both mathematics methods and content courses *con ganas*. He uses technology, multiple approaches and concrete materials to develop conceptual understanding of mathematical ideas. He has published over 120 articles about the teaching and learning of Mathematics. He has conducted workshops for students ranging from Kindergarten to College, and taught courses for pre-service and in-service teachers in 32 states in two countries.